BIOETHICS IN COMPLEXITY

FOUNDATIONS AND EVOLUTIONS

BIOETHICS
C⬤MPLEXITY
FOUNDATIONS AND EVOLUTIONS

editors

Sergio De Risio
Catholic University, Italy

Franco F Orsucci
Catholic University, Italy
&
Institute for Complexity Studies, Italy

ICP   Imperial College Press

Published by

Imperial College Press
57 Shelton Street
Covent Garden
London WC2H 9HE

Distributed by

World Scientific Publishing Co. Pte. Ltd.
5 Toh Tuck Link, Singapore 596224
USA office: Suite 202, 1060 Main Street, River Edge, NJ 07661
UK office: 57 Shelton Street, Covent Garden, London WC2H 9HE

British Library Cataloguing-in-Publication Data
A catalogue record for this book is available from the British Library.

BIOETHICS IN COMPLEXITY
Foundations and Evolutions

ISBN   1-86094-399-3

Printed in Singapore by World Scientific Printers (S) Pte Ltd

PREFACE

The primary motivation for writing and collecting this book is the awareness that bioethics, while trying to follow the fast developments of applied research, risks losing the connection with its own roots.

The very roots of ethics are immersed in the rich *humus* sedimented during centuries of development in Western thinking. This *humus* was named *Philosophia Naturalis*, Natural Philosophy, until its relatively recent splitting into many disciplines. The result of this division, due to a positivistic approach, has been an improvement of knowledge definition, while losing the integration of perspectives and comprehension of the same objects.

The main goal of this book is to recall the fundamentals of ethics in its dialogue with mind and society as embodied entities: recovering *Philosophia Naturalis* is a necessary premise for any bioethics. Without this integrative framework, bioethics risks being just the *ancilla* of applied research.

The result of this revitalization, under the guidance of masters, is a temporary healthy detachment from every-day news about new amazing discoveries in biotechnologies and genome trickeries.

We appreciate the thrill of a slow descent into the depths of Western Thinking. We feel the confusion of not finding immediate responses and answers. We experience the negative power of meeting questions, instead of answers.

San Juan de la Cruz used to say that every illumination must be preceded by darkness and ignorance. Freud, some centuries later, wrote in a letter to Lou Andreas Salome: "Sometimes I must artificially blind myself to converge all light on a dark spot."

The contributions in this book request a momentary *lapse* from current bioethical reasoning of public commissions and media activities. They re-pose the fundamental questions and suggest that ethics is born with interrogations on human nature and dialogues (*diàlogoi*) on the nature of actions.

Sergio De Risio, Franco F. Orsucci

Acknowledgments

We are grateful to Gabriella Frescura, Lakshmi Narayan and Ian Seldrup for their collaboration during the editorial phases in the development of this book.

Laura Monti and Maria Giovanna De Risio provided important assistance and suggestions.

Contents

Contents

[1]

ON A POSSIBLE FOUNDATION OF ETHICS

Sergio De Risio and Chiara Cuomo

Institute of Psychiatry and Clinical Psychology
Catholic University of Rome, Italy
psichiatria@rm.unicatt.it

Among the numerous and often interesting attempts to create a foundation upon which we could establish the principles of Ethics, two apparently almost incompatible perspectives exist, which nevertheless give us the cue to operate a convergence–operation that transforms them both and whose result could be that foundation we are looking for. We shall define them as the following:

- The perspective of a free descent of Ethics from above, and
- The perspective of a slow ascent of Ethics from below.

We refer to two great authors in depicting such perspectives, respectively K.-O. Apel for the first, and D. Dennett for the latter.

The path taken by Apel consists of a re-assumption of Kant's themes of Practical Reason—re-assumption that is a rotation of such themes—aimed at updating them to meet the requirements of contemporary living—around a definite anthropological, psychological and linguistic axis, which has been accumulating and consolidating during the 19th century.

Apel's Kantian re-assumption thus achieves an absolute originality, defined as *Discourse Ethics* (2001).

Since he wrote the *Fundamental Principals of the Metaphysics of Morals*, Kant (1998) has outlined the reference frame for a *material* discipline that most precisely corresponds to Ethics: the *moral philosophy*, intended as *science of freedom*.

In this frame, he enucleated the famous *categorical imperative*, formulated as: "Act always on such a maxim as thou canst at the same time will to be a universal law" or "Act on maxims which can at the same time have for their object themselves as universal laws of nature."

Kant then explained that the "three modes of presenting the principle of morality" are "so many formulae of the very same law." Such formulae, or forms of the same law, were adapted in order "to bring an idea of the reason nearer to intuition (by means of a certain analogy) and thereby nearer to feeling." They thus express that all the above maxims have:

1. "A form, consisting in universality (. . .)" (Kant, 1998).
2. A matter, namely, an end; and here the formula says that the rational being, as it is an end by its own nature and therefore an end in itself, must in every maxim

serve as the condition limiting all merely relative and arbitrary ends.

3. A complete characterization of all maxims by means of that formula, namely, that all maxims ought by their own legislation to harmonize with a possible kingdom of ends as with a kingdom of nature.

By bringing the second formula much closer to our notice (and always keeping in mind that this is the second of three formulae, which yet are forms of the very same law) it is possible to remark the prominence Kant gave to the rational element which also constitutes, in our opinion, the element of a natural continuity, having *The Critique of Pure Reason* as a basic position (1998).

The second chapter of *The Critique of Practical Reason* (Kant, 1998) entitled *Of the Concept of an Object of Pure Practical Reason*, actually begins by stating: "By a concept of the practical reason I understand the idea of an object as an effect possible to be produced through freedom (. . .) The only objects of practical reason are therefore those of good and evil. For by the former is meant an object necessarily desired according to a principle of reason; by the latter one necessarily shunned, also according to a principle of reason." (Kant, 1998)

This principle, this fundamental law of the *Pure Practical Reason*, had already been introduced by Kant in paragraph 7: "Act so that the maxim of thy will can always at the same time hold good as a principle of universal legislation" and considering it from a theorematic point of view, he adds the following corollary: "Pure reason is practical of itself alone and gives (to man) a universal law which we call the moral law."

We suppose it is possible to infer that what is shown by Kant is a sort of *inner voice*. A voice which arises from the *Pure Reason* every time that it meets in reality Good/Evil (*gut/böse*). A voice which every such time advises Will: "Shan't you wish for anything, under my direction (namely under the direction of me as Reason), unless you consider it *gut* or *böse*."

This implies that two characters are on stage: Reason and Will; and that we can hear a voice of the former which speaks to the latter. That author is Immanuel Kant.

This scene is enormously enlarged in K.O. Apel's *Discourse Ethics* (2001), where he keeps the *background music* of Rationality, but indefinitely increases the number of characters.

This transition (transformation in mathematical terms) operated by Apel on "the possible kingdom of ends," which is also the kingdom of *Pure Reason*, acquires reliable efficiency functional to—and because of—the knowledge accumulated upon the universe of *being in relation*. The major expression of such a sedimented knowledge is Freud's invention, based on that peculiar relationship of extraordinary heuristic significance, known as *psychoanalytic relationship.*

Apel does not always reveal a blind trust to psychoanalysis. He reproaches it as a discourse which does not occur between equals. However, as a matter of fact, in the analytic relationship the asymmetry of roles rests upon a deep and essential symmetry of subjects, which illustrates to the actors, with an absolute self-evidence and mutual acknowledgement, that they are of *equal dignity.*

Deprived of this necessary self-evidence, the analytical relationship is *not possible*. It is the frame that allows a *disclosure* according to the guidelines of freedom and rationality, which we have already learnt to know as close relatives of the ethic condition, or rather as a whole with the principle of ethics, defining the unique field where necessity and freedom coincide.

As a matter of fact, in the *psychotherapeutic discourses* (which are yet children of the psychoanalytic experience), Apel is nonetheless able to find a concrete model for the exercise of *Discourse Ethics*.

In fact, we could easily show that, within the psychotherapeutic discourse, an unethical attitude definitely implies the collapse of any possible therapeutic effect and in contrast obtains extremely harmful consequences.

Discourse Ethics is precisely concerned with the discussion of the cutting cdgc of pure rationality, before it faces any other *principle*.

This guarantees conclusions that will be better founded if:

1. the discourse will be developed according to reason;
2. subjects with equal dignity will be involved in the development of the discourse itself;
3. specific time will be devoted to discussion.

Discourse Ethics, because of these characteristics of a maxim, is not—and cannot—be restricted to the mere *regulation* of relationships among subjects or groups of subjects. To constitute a *model* of relations, i.e. among states (which as a whole conform themselves to a *possible kingdom of ends*), one must assume that the principle of rationality could take ubiquitous citizenship among those states.

Definitely, the *articulation* of the diplomatic system with the ideal forms of Discourse Ethics could assure in the future goals of peace and stability to each State as well as to all states as a whole.

We have tried to outline, with a certain conciseness, the crucial lines of thought for the Kantian categorical imperative to the Apel's Conversational Ethics.

We would like to recall, before referring to D. Dennett, that Ethics has always been conceived by human beings as an imperative coming from a superior being and expressing itself in rules whose trend, as evident in Apel, is the regulation of the human pragmatic aspect.

Whether this superior being's nature is religious, philosophical, or exquisitely mental (as in Freud's teaching) it is, in any case, a being of an elevated position, which addresses itself to the subject in order to bind him to a specific behavior or conduct.

When facing those rules embedded inside ourselves, that demand to be obeyed, we all feel a certain embarrassment, and annoyance. Although we sometimes are not deeply persuaded, we are nonetheless obliged to follow them.

A certain Conversational Ethics, already inspired by Pure Reason and supported by the rational discussion among purely equal human beings, could involve a certain higher degree of firm belief in this natural human being.

However, something transcendental, even too transcendental, remains in pure equal human beings. This is the reason which drives Dennett to make his brave attempt to build a bio-cultural Ethics, starting from his impassioned and impassioning studies on the field of the so-called Artificial Intelligence: an artifice according to nature and, more precisely, according to a rigorously and definitely Darwinian nature.

The most meaningful contribution in that field, due to Dennett (1995), a Professor who devotes himself to teaching with the same passion that appears in his research, is mostly in his *Darwin's dangerous idea.*

What could be ever so dangerous in an idea? We certainly know that various ideas involve pragmatic ruinous effects in the more or less metaphorical meanings of the word. This is why every idea, before being used, should certainly be taken cautiously: a kind of precautionary *filtering* applied to an idea before it is put into practice.

However, this is not the aim of *Darwin's dangerous idea*. In sketching out a form of Ethics *which arises from below*, Dennett emphasizes that an idea is dangerous solely with respect to another idea. The danger of an idea is thus, first of all, danger for other ideas, perhaps for those which have firmly stabilized themselves in one's mind, and which are neglected in that they are deprived of a relationship with other critical, competitive ideas: without competition, a real competence may never be developed.

Dennett names such 'in-competition-ideas' *memes*, recovering the term, coined by Dawkins in 1976 (1982), with the fusion of the names *gene* and *memory*. The *meme*, a cultural unit, tends to evolve in that specific direction which allows for optimal diffusion and survival. It thus transcends and goes *beyond good and evil*.

This inclination to spread is, in fact, the meme's primary feature, entirely independent of its value. Yet, if only bad memes, those harmful for their bearers, had diffused, we would not be here writing, reading, singing, painting and observing images, composing and listening to music nor would we be trying to inquire about memetics.

This is the reason for which the diffusion of memes cannot be regulated by a pure *mimetic* law. The pure self-regulating *mimesis* is analogous to the degree zero of Ethics. However, it prescriptively aims at the construction of what we can properly consider as an evolutionary Ethics, and this represents the inclination of Ethics to infinity.

Dennett writes that biology imposes some bonds to those we can consider as our values. They wouldn't survive in the long term if we were not used to choose the memes which are helpful to us, and if this choice did not yield better results than a random choice. However, we have not seen the long term yet. Mother Nature's cultural experiment on this planet has been in progress for some thousand generations only.

Nevertheless, this very condition does not elude a self-evident observation. We cannot avoid noticing, among all our *mind-dwellers*, that memes of normative concepts, such as truth, goodness, beauty or duty, are set up in the most stable way. We are therefore watching and waiting. Mother Nature's experiment is still continuing.

Among these steadily mind-anchored memes, there is one which is particularly diffused. A universal one, which Freud designated as the *Oedipus Complex* describing it as a set of rules applied on that instinct, so important to survival yet so whimsical, which is referred to as the sexual instinct.

We wonder whether Mother Nature may, throughout that endless progress which most properly belongs to her, elaborate a meme somehow analogous to the Oedipus Complex, which could thus provide the Will of Power with a set of rules. The experiment risks becoming particularly interesting. First of all, because it would be necessary, for a meme analogous to the Oedipus Complex, to settle in our minds in the most stable manner, so stable that it would be able to gain that universal significance assured by the Oedipus Complex. For, as it is well known, the Oedipus Complex is a structure stably anchored in the unconscious.

In the game of the endless multiple oppositions typifying the mind, and the relationships among minds, only a symbolic principle will be able to ensure the possibility of *knowing-with*, *feeling-with*, and *coming-with* for the abstract foundation of that ἔθος which is παλύντροπος ἁρμοζεῖν.

In the meantime, let us ask ourselves what is basically universal in the Freudian concept. Obviously universal is the symbolic significance of this Oedipus Complex, symbolic significance which is most evident in both Jacques Lacan's *Écrits*, particularly in *Kant with Sade* and in his Seminar dedicated to the *Ethics of Psychoanalysis*. This Seminar essentially deals with the relationship between *la Loi* (the Law), the Name of the Father, the Symbolic Other, and the laws (or *lawlets*) that take you to bed (or couch) on the basis of a misunderstood right to pleasure.

These daily lawlets, often written with the traits of love letters, gradually reveal themselves with their true nature, which is that of sadistic letters. Exceptions to the rule derail the subjects, withdrawing them from the reasonable exercise of Conversational Ethics as well as from every possible evolution following a reasonable πόλεμος (*pòlemos*), in the Heraclitean sense. Πόλεμος, which we prefer to translate with a liberal word: *competition*.

Let us remember what we have already said, but what is never sufficiently repeated: without competition, a real, proper and full competence may never be achieved.

[2]

In Darwin's Wake, Where Did I Go?

Daniel C. Dennett

Center for Cognitive Studies
Tufts University, Boston, MA, USA
ddennett@tufts.edu

> *Parfois je pense; et parfois, je suis.*
> Paul Valéry

> *Je ne cherche pas; je trouve.*
> Pablo Picasso

1. Valéry's "*Variation sur Descartes*" (in 1974) excellently evokes the vanishing act that has haunted philosophy ever since Darwin overturned the Cartesian tradition: the incredible Disappearing Self (Dennett, 1984). One of Darwin's earliest critics saw what was coming and could scarcely contain his outrage: "In the theory with which we have to deal, Absolute Ignorance is the artificer; so that we may enunciate as the fundamental principle of the whole system, that, IN ORDER TO MAKE A PERFECT AND BEAUTIFUL MACHINE, IT IS NOT REQUISITE TO KNOW HOW TO MAKE IT. This proposition will be found, on careful examination, to express, in condensed form, the essential purport of the Theory, and to express in a few words all Mr. Darwin's meaning; who, by a strange inversion of reasoning, seems to think Absolute Ignorance fully qualified to take the place of Absolute Wisdom in all the achievements of creative skill." (MacKenzie, 1868 cit. in Dennett, 1984)

2. This "strange inversion of reasoning" promises—or threatens—to dissolve the Cartesian *res cogitans* as the wellspring of creativity, and then where will we be? Nowhere, it seems. It seems that if creativity gets "reduced" to "mere mechanism" we will be shown not to exist at all. Or, we will exist, but we won't be thinkers, we won't manifest genuine "Wisdom in all the achievements of creative skill." Whenever we zoom in on the act of creation, it seems we lose sight of it, the intelligence or genius replaced at the last instant by stupid machinery, an echo of Darwin's shocking substitution of Absolute Ignorance for Absolute Wisdom in the creation of the biosphere. How can I be a moral agent, a responsible author of my own acts (and not merely a shifting nexus in the great fabric of blind causation), if my soul is replaced with an evolved mechanism? Many people dislike Darwinism in their guts, and of all the ill-lit, murky reasons for antipathy to Darwinism, this one has always struck me as the deepest, but only in the sense of being the most entrenched, least accessible to rational criticism. There are thoughtful people who scoff at creationism, dismiss dualism out of hand, pledge allegiance to academic humanism—and then get quite nervous when it is suggested that a Darwinian theory of creative intelligence might be in the cards, and might demonstrate that all the works of human genius can be understood in the end to be products of a cascade of generate-and-test procedures that are, at bottom, algorithmic, mindless. Absolute Ignorance? Artificial Intelligence? Fie on anybody who would thus put "A" and "I" together!

3. Unsupportable antipathies often survive thanks to protective coloration: they blend into the background of legitimate objections to overstatements of the view under attack. Since the reach of Darwinian enthusiasm has always exceeded its grasp, there are always good criticisms of Darwinian excesses to hide amongst. And so the battle rages, generating as much suspicion as insight. Today I want to examine the question of the biopsychological roots of ethics by an indirect method: by looking at the more tractable, simpler, problem of the biopsychological roots of artistic creativity. The two issues are closely united in any case: we want to be the authors of both our moral deeds and our artistic creations; we want to be responsible in both cases, and it is this responsibility that appears to be jeopardized by the Darwinian vision.

4. Darwinians who are sure that a properly nuanced, sophisticated Darwinism is proof against all the objections and misgivings—I am one such—should nevertheless recall the fate of the Freudian nags of the 50s and 60s, who insisted on seeing everything through the perspective of their hero's categories, only to discover that by the time you've attenuated your Freudianism to accommodate everything, it is Pickwickian most of the way. Sometimes a cigar is just a cigar, and sometimes an idea is just an idea—not a meme—and sometimes a bit of mental machinery is not usefully interpreted as an adaptation dating back to our ancestral hunter-gatherer days or long before, even though it is, obviously, descended (with modifications) from some combination or other of such adaptations. We Darwinians will try to remind ourselves of this, hoping our doughty opponents will come to recognize that a Darwinian theory of creativity is not just a promising solution but the only solution in sight to a problem that is everybody's problem: how can an arrangement of a hundred billion mindless neurons compose a creative mind?

5. William Poundstone, in his elegant 1985 book about Conway's game of Life, *The Recursive Universe: Cosmic Complexity and the Limits of Scientific Knowledge*, puts the inescapable challenge succinctly in terms of "the old fantasy of a monkey typing Hamlet by accident." He calculates that the chances of this happening are "1 in 50 multiplied by itself 150,000 times."

6. In view of this, it may seem remarkable that anything as complex as a text of Hamlet exists. The observation that Hamlet was written by Shakespeare and not some random agency only transfers the problem. Shakespeare, like everything else in the world, must have arisen (ultimately) from a homogeneous early universe. Any way you look at it Hamlet is a product of that primeval chaos.

7. Where does all that Design come from? What processes could conceivably yield such improbable "achievements of creative skill"? What Darwin saw—and we might just as well couch it in economic terms, since they excellently express a deep truth in any age to anyone who will listen—is that Design is always both valuable and costly. It does not fall like manna from heaven, but must be accumulated the hard way, by time-consuming, energy-consuming processes of mindless search through "primeval chaos," automatically preserving happy accidents when they occur. This broadband process of Research and Development is breathtakingly inefficient, but—this is Darwin's great insight—if the costly fruits of R and D can be thriftily conserved, copied, stolen, and re-used, they can be accumulated over time to yield "the achievements of creative skill." "This principle of preservation, I have called, for the sake of brevity, Natural Selection."

8. There is no requirement in Darwin's vision that these R and D processes run everywhere and always at the same tempo, with the same (in-)efficiency. If we think of design work as lifting in Design Space (an extremely natural and oft-used metaphor, exploited in models of hill-climbing and peaks in adaptive landscapes, to name the most obvious and popular applications), then we can see that the gradualist, frequently back-sliding, maximally inefficient basic search process can on important occasions yield new conditions that speed up the process, permitting faster, more effective local lifting. Call any such product of earlier R and D a crane, and distinguish it from what Darwinism says does not happen: skyhooks. Skyhooks, like manna from heaven, would be miracles, and if we posit a skyhook anywhere in our "explanation" of creativity, we have in fact conceded defeat—"Then a miracle occurs." What, then, is a mind? Picasso's self-description (which I do not believe for one instant) excellently evokes the extreme: it is a Finder that need not Search, a Designer of such genius that it can finesse all the stupid lifting work and simply descend from on high, miraculously, on one of the beauty spots, one of the peaks in Design Space.

9. We can now characterize the mutual suspicion between Darwinians and anti-Darwinians that distorts the empirical investigation of creativity. Darwinians suspect their opponents of hankering after a skyhook, a miraculous gift of Genius whose powers have no decomposition into mechanical operations, however complex and informed by earlier processes of R and D. Anti-Darwinians suspect their opponents of hankering after an account of creative processes that so diminishes the Finder that it disappears, at best a mere temporary locus of mindless differential replication. We can further illuminate the anti-Darwinians' concern by adapting Poundstone's example of the creation of the creator of Hamlet.

10. If Dr Frankenstein designs and constructs a monster, Spakesheare, molecule by molecule, that thereupon sits up and writes out a play, Spamlet, then Dr Frankenstein is probably the author of Spamlet, using his intermediate creation, Spakesheare, as a mere storage-and-delivery device, a fancy word processor of sorts. How could we tell? We look inside Spakesheare. If we find Spamlet, chapter and verse, already composed and ready to deliver to the hand-writer module, it is obvious. If we find instead that Spakesheare has been equipped by Dr Frankenstein with a virtual past, a lifetime stock of pseudo-memories of experiences on which to draw while responding to its Frankenstein-installed obsessive desire to write a play, we will see the processes that actually occur inside Spakesheare, after it becomes animated, as playing a significant creative role. After all, they will involve multi-leveled searches, guided by multi-leveled evaluations (including evaluation of the evaluation . . . of the evaluation functions as a response to evaluation of . . . the products of the ongoing searches). We may be curious about whether the amazing Dr Frankenstein actually anticipated all this activity and designed Spakesheare's virtual past (and all the search machinery) to yield just this product, Spamlet. If so, then Dr Frankenstein is surely the author of

Spamlet, as well as being the author of Spakesheare. But if Dr Frankenstein is simply unable to foresee all this (a smidgen of realism in our wild fantasy), and is delegating to Spakesheare all the hard work (the way Deep Blue's creators delegated to it the hard work of finding winning trajectories against Kasparov), then Spakesheare itself is undoubtedly the locus of the creativity, whether or not we find it in our hearts to grant Spakesheare moral status as responsible for authoring Spamlet. Nobody else wrote Spamlet (nobody else beat Kasparov), so either we get used to artifactual authors, or we get used to authorless plays.

11. I find it interesting that some anti-Darwinians find this choice point to be their main chance, the crack into which they may best hope to drive a wedge, ensuring a gulf separating Art (and Mind) from the rest of Nature. According to this vision, an artifact, a robot, could never be a genuine Author of acts in the sense that is required for being morally responsible for those acts. The trouble with this strategy is that it is unusually vulnerable to penumbral cases that threaten to span the gap—not just imaginable penumbral cases, but real ones. In the realm of music, the Frankenstein-Spakesheare fantasy is already a reality, in the person(s) of David Cope and EMI, his creation and creator in turn of some thousands of estimable musical compositions. Cope set out to create a mere efficiency-enhancer, a composer's aid to help him over the blockades of composition any creator confronts, a high-tech extension of the traditional search vehicles (the piano, staff paper, the tape recorder, etc.). As EMI grew in competence, it promoted itself into a whole composer, incorporating more and more of the generate-and-test process. David Cope can no more claim to be the composer of EMI's symphonies and motets and art songs than Murray

Campbell can claim to have beaten Kasparov in chess. To a Darwinian, this new element in the cascade of cranes is simply the latest in a long history, and we should recognize that the boundary between authors and their artifacts should be just as penetrable as all the other boundaries in the cascade. When Richard Dawkins notes that the beaver's dam is as much a part of the beaver phenotype—its extended phenotype—as its teeth and its fur, he sets the stage for the further observation that the boundaries of a human author are exactly as amenable to extension. In fact, of course, we've known this for centuries, and have carpentered various semi-stable conventions for dealing with the products of Michelangelo, of Michelangelo's studio, of Michelangelo's various pupils.

12. Wherever there can be a helping hand, we can raise the question of just who is helping whom, what is creator and what is creation. How should we deal with such questions? To the extent that anti-Darwinians simply want us to preserve some tradition of authorship, to have some rules of thumb for determining who or what shall receive the honor (or blame) that attends authorship, their desires can be acknowledged and met, one way or another (which doesn't necessarily mean we should meet them). To the extent that this is not enough for the anti-Darwinians, to the extent that they want to hold out for authors as an objective, metaphysically grounded, "natural kind" (oh, the irony in those essentialist wolf-words in naturalist sheep's clothing!), they are looking for a skyhook.

13. This renunciation of skyhooks is, I think, the deepest and most important legacy of Darwin in philosophy, and it has a huge domain of influence, extending far beyond the skirmishes of evolutionary epistemology and evolutionary ethics. If we commit ourselves to Darwin's "strange inversion of reasoning," we turn our backs on compelling ideas that have been central to the philosophical tradition for centuries, not just Aristotle's essentialism and irreducible telos, but also Descartes's *res cogitans* as a causer outside the mechanistic world, to name the three that had been most irresistible until Darwin came along. The siren songs of these compelling traditions still move many philosophers who have not yet seen fit to execute the inversion, sad to say. Clinging to their pre-Darwinian assumptions, they create problems for themselves that will no doubt occupy many philosophers for years to come. The themes all converge when the topic is creativity and authorship, where the urge is to hunt for an "essence" of creativity, an "intrinsic" source of meaning and purpose, a locus of responsibility somehow insulated from the causal fabric in which it is embedded, so that within its boundaries it can generate, from its own genius, its irreducible genius, the meaningful words and deeds that distinguish us so sharply from mere mechanisms.

14. Plato called for us to carve Nature at its joints, a wonderful biological image, and Darwin showed us that the salient boundaries in the biosphere are not the crisp set-theoretic boundaries of essentialism, but the emergent effects of historical processes. As one species turns into two, the narrow isthmus of intermediates evaporates as time passes, leaving islands, concentrations sharing family resemblances, surrounded by empty space. As Darwin noted, in somewhat different terms, there are feedback processes that enhance separation, depopulating the middle ground. We should expect the same sort of effects in the sphere of human mind and culture. Do products of the processes of artistic creation typically spring from a single mind? "Are you the author of this?" "Is this all your own work?" The mere fact that these are familiar questions shows that there are strong cultural pressures encouraging people to make the favored answers come true. A small child, crayon in hand, huddled over her drawing, slaps away the helping hand of parent or sibling, because she wants this to be her drawing. She already appreciates the norm of pride of authorship, a culturally imbued bias built on the palimpsest of territoriality and biological ownership. The very idea of being an artist shapes her consideration of opportunities on offer, shapes her evaluation of features she discovers in herself. And this in turn will strongly influence the way she conducts her own searches through design space, in her largely unconscious emulation of Picasso's ideal, or, if she is of a contrarian spirit, defying it, like Marcel Duchamp:

> Cabanne: What determined your choice of *readymades*?
> Duchamp: That depended on the object. In general, I had to beware of its "look." It's very difficult to choose an object, because, at the end of fifteen days, you begin to like it or to hate it. You have to approach something with an indifference, as if you had no aesthetic emotion. The choice of *readymades* is always based on visual indifference and, at the same time, on the total absence of good or bad taste. . . .

15. Since the tubes of paint used by the artist are manufactured and ready made products we must conclude that all the paintings in the world are "readymades aided" and also works of assemblage.

16. There is a persistent problem of imagination management in the debates surrounding this issue: people on both sides have a tendency to underestimate the resources of Darwinism, imagining simplistic alternatives that do not exhaust the space of possibilities. Darwinians are notoriously quick to find (or invent) differences in fitness to go with every difference they observe, for instance, while anti-Darwinians, noting the huge distance between a bee-hive and the St. Matthew Passion as created objects, are apt to suppose that anybody who proposes to explain both creative processes with a single set of principles must be guilty of one reductionist fantasy or another: "Bach had a gene for writing baroque counterpoint just like the bees' gene for forming wax hexagons" or "Bach was just a mind-less trial-and-error mutator and selector of the musical memes that already flourished in his cultural environ-ment." Both of these alternatives are patent nonsense, of course, but pointing out their flaws does nothing to sup-port the idea that ("therefore") there must be irreducibly non-Darwinian principles at work in any account of Bach's creativity. In place of this dimly imagined chasm with "Darwinian phenomena" on one side and "non-Darwinian phenomena" on the other side, we need to learn to see the space between bee and Bach as populated with all man-ner of mixed cases, differing from their nearest neighbors in barely perceptible ways, replacing the chasm with a traversable gradient of non-minds, protominds, hemi-demi-semi minds, copycat minds, aping minds, magpie minds, clever-pastiche minds, and eventually, genius minds. And the individual minds, of each caliber, will themselves be composed of different sorts of parts, including, surely, some special-purpose "modules" adapted to various new tricks and tasks, as well as a cascade of higher-order reflection devices, capable of generating ever more rarefied and de-limited searches through pre-selected regions of the Vast space of possible designs. At no point as we traverse this gradient do we "desert the Darwinian framework entirely, and . . . engage, head-on, with all the serious epistemological problems of understanding methods of discovery," as Kitcher puts it (1997) in an uncharacteristic over-statement.

17. Darwin himself shows us how to unite the range of possibilities into a single perspective. In the first chapter of *Origin of Species*, Darwin (1997) introduces his idea of natural selection by an ingenious expository device, an instance of the very gradualism that he was about to discuss. He begins not with natural selection—his destination—but what he calls methodical selection: the deliberate, foresighted, intended "improvement of the breed" by animal and plant breeders. He begins, in short, with familiar and uncontroversial ground that he can expect his readers to share with him.

18. We cannot suppose that all the breeds were suddenly produced as perfect and as useful as we now see them; indeed, in several cases, we know that this has not been their history. The key is man's power of accumulative selection: nature gives successive variations; man adds them up in certain directions useful to him.

19. But, he goes on to note, in addition to such methodical selection, there is another process, which lacks the foresight and intention, which he calls unconscious selection:

> At the present time, eminent breeders try by methodical selection, with a distinct object in view, to make a new strain or sub-breed, superior to anything existing in the country. But, for our purpose, a kind of Selection, which may be called Unconscious, and which results from every one trying to possess and breed from the best individual animals, is more important. Thus, a man who intends keeping pointers naturally tries to get as good dogs as he can, and afterwards breeds from his own best dogs, but he has no wish or expectation of permanently altering the breed. (Darwin, 1977)

20. Long before there was deliberate breeding, unconscious selection was the process that created and refined all our domesticated species, and even at the present time, unconscious selection continues. Darwin gives a famous example: "There is reason to believe that King Charles's spaniel has been unconsciously modified to a large extent since the time of that monarch." (Darwin, 1977)

21. There is no doubt that unconscious selection has been a major force in the evolution of domesticated species. In our own time, unconscious selection goes on apace, and we ignore it at our peril. Unconscious selection in bacteria and viruses for resistance to antibiotics is only the most notorious and important example. Darwin pointed out that the line between unconscious and methodical selection was itself a fuzzy, gradual boundary: "The man who first selected a pigeon with a slightly larger tail, never dreamed what the descendants of that pigeon would become through long-continued, partly unconscious and partly methodical selection." (Darwin, 1977)

22. And both unconscious and methodical selection, he notes finally, are not alternatives to, but special cases of, a more inclusive process, natural selection. From this inclusive perspective, changes in lineages due to unconscious or methodical selection are simply instances of natural selection in which one of the most prominent selective pressures in the environment is human activity.

23. This nesting of different processes of natural selection now has a new member: genetic engineering. How does it differ from the methodical selection of Darwin's day? It is less dependent on the pre-existing variation in the gene pool, and proceeds more directly to new candidate genomes, with less overt trial and error. Darwin had noted that in his day: "Man can hardly select, or only with much difficulty, any deviation of structure excepting such as is externally visible; and indeed he rarely cares for what is internal." (Darwin, 1977)

24. But today's genetic engineers have carried their insight into the molecular innards of the organisms they are trying to create. There is ever more accurate foresight, but even here, if we look closely at the practices in the laboratory, we will find a large measure of exploratory trial and error in their search of the best combinations of genes.

Are the products of genetic engineering "Darwinian" products? They are produced not by blind or random trial-and-error variation, but by highly intelligent, guided, foresightful processes. Nevertheless these processes are themselves the products of earlier design work accomplished by Darwinian R and D, and if we look closely at the microprocesses that compose their current, local search, we will still find plenty of instances of random (undesigned, chaotic) generation of candidates for further scrutiny.

25. It may seem, however, that we have now passed the *Pickwickian* limits of Darwinian orthodoxy. Does a Darwinian gloss actually supplement or adjust the traditional intellectualist ways of thinking? I think it does, because without the steady pressure of the Darwinian "strange inversion of reasoning," it is all too tempting to revert to the old essentialist, Cartesian perspectives. For instance, there is always the temptation, often succumbed, to establish *principled* boundaries, or to erect a polar contrast between insightful and blind processes of search. This is particularly evident in the frequent observation that whereas Deep Blue beat Kasparov at chess, its brute force search methods were entirely unlike the exploratory processes Kasparov used. That is simply not so—or at least the Darwinian perspective warns us that such a claim, in this post-Darwinian age, is achingly in need of support, with none in sight. Kasparov's brain is made of organic materials, and has an architecture importantly unlike that of Deep Blue, but it is still, so far as we know, a massively parallel search engine which has built up, over time, an outstanding array of heuristic pruning techniques that keep it from wasting time on unlikely branches. There is no doubt that Kasparov is the beneficiary of a cascade of cranes that have extracted good design principles from past games—and from the lives of Kasparov's ancestors, and many other sources contributory to the culture with which Kasparov has been imbued since childhood—so that he can recognize, and know enough to ignore, huge portions of the game space that Deep Blue must still patiently canvass. Kasparov's *insight*, the product of R and D not yet available to Deep Blue, dramatically focuses the search for Kasparov, but it no more constitutes an exception to Darwinian principles than the methodical breeding of domesticated animals constitutes a counterexample marring the universality of Darwinian evolution by natural selection.

26. It is important to recognize that genius is itself a product of natural selection and involves generate-and-test procedures all the way down. Once you have such a product, it is often no longer particularly perspicuous to view it solely as a cascade of generate-and-test processes. It often makes good sense to leap ahead on a narrative course, thinking of the agent as a self, with a variety of projects, goals, presuppositions, hopes, ... In short, it often makes good sense to adopt the intentional stance towards the whole complex product of evolutionary processes, putting the largely unknown and unknowable mechanical microprocesses and the history that set them up out of focus and highlighting the patterns of rational activity that those mechanical microprocesses track so closely. It especially makes good sense to the creator himself or herself, who must learn not to be oppressed by the revelation that on close inspection, even on close introspection, a genius dissolves into a pack rat, which dissolves in turn into a collection of trial-and-error processes over which nobody has ultimate control.

27. Does realization amount to a loss—an elimination—of selfhood, of genius, of creativity? Those who are closest to the issue—the artistic and scientific geniuses who have reflected on it—offer sharply different opinions, but I think that this is due more to their differing susceptibility to the pre-Darwinian myths than anything they have privileged access to in the privacy of their studios, their minds.

28. When the mathematician Henri Poincaré (1929), reflected on this topic, he saw only two alternatives and they were both disheartening to him. The unconscious self that generates the candidates "is capable of discernment; it has tact, delicacy; it knows how to choose, to divine. What do I say? It knows better how to divine than the conscious self since it succeeds where that has failed. In a word, is not the subliminal self superior to the conscious self? I confess that, for my part, I should hate to accept this."

29. Mozart was less disheartened; he is reputed to have said of his best musical ideas: "Whence and how do they come? I don't know and I have nothing to do with it." The painter Philip Guston (2001) is equally unperturbed by this evaporation of visible self when the creative juices start flowing: "When I first come into the studio to work, there is this noisy crowd which follows me there; it includes all of the important painters in history, all of my contemporaries, all the art critics, etc. As I become involved in the work, one by one, they all leave. If I'm lucky, every one of them will disappear. If I'm really lucky, I will too."

30. And the novelist and essayist, Nicholson Baker, has this to say on the way we go about changing our minds: "Our opinions, gently nudged by circumstance, revise themselves under cover of inattention. We tell them, in a steady voice, 'No, I'm not interested in a change at present.' But there is no stopping opinions. They don't care about whether we want to hold them or not; they do what they have to do."

31. Darwin saw that the phenomena of methodical and unconscious selection were not counterexamples to his principle of natural selection because the very methodical (conscious or unconscious) selectors were themselves products of earlier rounds of natural selection, and hence just very, very specialized selective environments in which differential replication could occur. Within their enclosures, isolated domesticated subpopulations competed, willy nilly, for replicative chances, and any imagined distinction between conscious and unconscious imposition of the determining selection pressures made no systematic difference to the outcome. Thoughts, ideas both artistic and scientific, opinions, similarly compete for chances to recur, to spawn elaborations of themselves in private and public. Can a mind, a self, be constructed out of such processes? Darwin showed us how the answer could well be Yes, and no non-miraculous alternatives have ever been formulated.

Thanks to Nicholas Humphrey, Will Lowe and Victoria McGeer for ideas that helped me improve an earlier draft of this essay.

[3]

Post-Kantian Problems of an Ethics of the Good Life and the Foundations of Discourse Ethics

Karl-Otto Apel

Department of Philosophy, Goethe-Universität
Frankfurt, Germany

1 The Choice of My Topic Justified

My attempts at unfolding my approach of a dialogical or discourse ethics in the last decades have primarily developed into the direction of an (universal) ethics of justice and (global) co-responsibility, i.e. of a social and political macro-ethics with a growing affinity to the institutions of democracy, law and finally even economy. Now, I feel confronted with a difference and tension between two current perspectives and approaches to the problem of ethics: namely between the perspectives of utilitarian and Kantian types of ethics, on the one hand, and those of Aristotelian or, nowadays, existentialist types of an ethics of the good life or self-realization, on the other hand. And it seems obvious to me that the latter perspective is more closely connected with the professional concern of psychiatry than the former.

In order to respond to this situation, I have decided to write about this very difference and tension between current approaches to ethics from the point of view of discourse ethics. Hence I have chosen the topic "Post-Kantian problems of an ethics of the good life and the foundation of discourse ethics."

2 Ethics of the Good Life and other Problems of Ethics after Kant's Innovation of Universalistic Ethics

In order to win a provocative starting point for my chapter let me take recourse to Michael Foucault's last public inter-view in 1984. Already in his first greater work, *Histoire de la folie* (1973), Foucault had conceived of the rise of modern reason as "negation, expulsion and reduction" of all those features of culture that were to be considered as *extériorité* from the point of view of one universal norm of reason. Correspondingly in his last work, *Histoire de la sexualité* (1978), Foucault regarded the normative ethics of universal principles, which emerged with Stoicism and Christianity and developed further with Kant, as the worst enemy of the old Greek ethics of the *souci de soi*, i.e. the striving for a beautiful way or style of self-realization. And in his interview of 1984, Foucault stated: «la recherche d'une forme morale qui serait acceptable par tout le monde en ce sens que tous devait s'y soumettre me parait catastrophique.»

Here we find the sharpest expression of that contrast between the current perspectives on ethics I mentioned at the beginning. And it also becomes clear from Foucault's position that it is possible to conceive of the modern post-existentialist ethics of authentic self-realization as an approach that can at least account for its own concern by taking recourse to the ancient, especially the Aristotelian ethics of the good life.

There is, though, a point of the Foucaultian polemics that is foreign to the classical Greek conception of the good life. Although Plato's and Aristotle's *polis*-related ethics was not yet universalistic in a cosmopolitan sense, as was, for the first time, the Stoic and the Christian ethics, nevertheless the universalistic concern of justice was implied in the Greek conception of the good life. This becomes indirectly clear from the fact that the Greek classics after all found it necessary to give a reason for the three well known deviations of their ethics from universalism, namely the minor rights attributed to the barbarians, the slaves and the women. The reason for these deviations was given by the argument that the barbarians, the so-called born slaves, and the women have a lesser share in (of) the logos.

By contrast, Foucault and similar pleas for an ethics of self-realization in our time are polemically directed against the idea of universalism as such. This raises the need for a special explanation. It can be provided, in my opinion, by an account of the Kantian innovation in the development of universalistic ethics of justice: an innovation that definitely separated this ethics from the ancient conception of an ethics of the good life (or of *eudaimonia*). It even introduced an other supreme standard or yardstick of the good.

For Kant there is no telos of my good life or happiness that could provide the decisive yardstick of morality, i.e. of a universally valid orientation of the "good will" in following maxims of actions. This does not mean that Kantian ethics would forbid or exclude the individual persons' striving for happiness. On the contrary, it sets it free as a right of everybody; but at the same time it subjects it to the principle of its compatibility with the moral law, i.e. with the universalization of the maxims of one's action as a possible law for all possible actors. And this universalization principle, and not the *souci de soi* as striving for one's individual happiness, with Kant becomes the supreme normative principle of ethics.

Now, this innovation deeply cut into the development of occidental ethics, changing even the tendencially universal ethics of justice that had been already initiated by the Stoic doctrine of natural rights and by Christianity. For it distinguished a normative yardstick of the good as the right that was no longer contrasted to the bad as related to a valuation of individual (or even collective, social) preferences; being related to the unconditional demand of the universally valid moral law, it was rather contrasted to the evil as negation of the law of moral reason. Thus far Kant's innovation was the origin of a deontological ethics. It was no longer dependent on valuations of the good or the bad for some purpose or to some particular human subject, but, according to Kant, it was corresponding to the "autonomous," "legislative" will of the subject of reason. As subject of practical reason it was a member or co-subject with equal rights and duties within the "realm of purposes" (which, according to Kant, includes human beings as purposes in themselves together with God).

Now, this metaphysical foundation of a deontological ethics created a variety of resultant novel problems. For, Kant's distinction between unconditioned "duty" and empirical "inclinations" indeed provides a negative principle for an examination of possible maxims of actions through pointing to an unconditional measure of rightness. But it cannot really serve as a sufficient method of positive orientation with regard to those problems of practical reason that have to be solved in the context of experience in the living world of individuals or groups.

Thus already the experiment of thought that is imposed on the individual actor by Kant's "categorical imperative," in order to enable him or her to find out whether a possible maxim of action is acceptable as a universal law, in many cases overtaxes the individual competence; or it suggests to the individual solutions that are only universalizing, if he/she already presupposes traditional conventions or institutions as an unquestionable matter of course. Hegel, from this point of view, called the applicability of the categorical imperative into question. Thus he pointed to the fact that in the case of Kant's example of the *depositum* that must not be embezzled, the institution of private property is presupposed, and in the same sense, according to Hegel, every "ought" of Kantian morality presupposes already the reality of "substantial morality" (*substantielle Sittlichkeit*) on the level of the "objective spirit" of human culture. Therefore Hegel replaced the unconditioned and unhistorical yardstick of the Kantian principle of universalization by the supposition that the universally valid morality of freedom and reason will necessarily realize itself in the course of history, i.e. through the sequence of the national bearers of the substantial morality of the "objective spirit."

It is well known that in our day the so-called "communitarian" philosophers have practically renewed the Hegelian conception of culture- and community-dependent "substantial morality"; but, of course, these modern, Anglo-Saxon philosophers, like e.g. Ch. Taylor, Sandel, A. MacIntyre, M. Walzer and others, renewed the Hegelian conception without its internal connection with a metaphysics of history; and this meant that the historicist relativism of culture-dependence, which with Hegel was veiled (concealed) by the metaphysics of historical progress, now emerged as a serious (grave) aporia in the face of the problems of inter-cultural justice and co-responsibility in an age of globalization and multiculturalism. In fact, the Hegelian discovery of the culture- and community-dependence of "substantial morality," which was in our time renewed and deepened by hermeneutic philosophy and by communitarianism, cannot be ignored; but it now turns out that the Kantian principle of universalization, conceived as a non-culture-dependent yardstick of examination, even and precisely in the face of different and competing traditions of substantial morality, cannot be dispensed with either. Rather, a novel and urgent problem of mediation in ethics turns up here. I shall come back to this later.

Another problem of practical orientation that can be understood as a dialectical consequence of the Kantian foundation of deontology, is indicated by terms like utilitarianism, consequentialism, and ethics of responsibility put against an ethics of good intentions (*Gesinnungsethik* in Max Weber's sense). Kant, it is true, has not overlooked the structure and importance of means-ends-rationality that is essentially pre-supposed in all of these positions that are opposed to pure deontologism. But Kant restricted the scope of this rationality to that of "hypothetical imperatives" which he understood as "pragmatical advices of prudence" which are morally neutral since they are essentially different from the unconditional demands of the "categorical imperative."

Now it is true that in all of these contra-deontological positions there remains an aporia of the ultimate moral yard-stick, so to speak. Thus pure utilitarianism cannot say what the use that is to be maximized is to be good (i.e. useful) for, and, without a supplementation by a universalization principle, utilitarianism is always exposed to the temptation of maximizing the use for a collective at the expense of a minority, i.e. by a violation of justice. In a similar sense, pure consequentialism lacks a last yard-stick of judging the possible consequences of our actions, and, besides that, it is practically impossible in many cases to win sufficient knowledge of all possible consequences of action. This last argument is also pertinent to an ethics of responsibility; and, beyond that, the will to take over responsibility for all recognizable consequences of an action makes a person a possible victim of blackmailing (extortion). Hence, even M. Weber was prepared to admit that even a responsible politician can be compelled to give preference to an ethics of the good intention by saying to himself: I will not do this and this, on principle, whatever the consequences might be.

Nevertheless, there is no doubt that, in the context of life praxis, there is a morally relevant need for orientation toward useful rather than harmful consequences of actions: a need that is not taken into account by Kant's strict deontologism. It is not true for example that looking for the consequences of our actions, if it is more than a morally neutral advice of prudence, is always looking for the useful consequences for the actor himself, as Kant seems to suppose—especially with regard to the motive of heavenly reward. And it seems to me untenable to give the advice—along with Kant—to the "moral politician" that he should act according to the principle: *fiat justitia, pereat mundus*. This lies of course along the same line as Kant's famous refusal of allowing a white lie to the person who is asked by a murderer for the stay (hiding place) of his victim in his house. Hence, also in this respect, i.e. with regard to our responsibility for the consequences of our actions, there remains an unsolved resulting problem of Kant's foundation of strict deontologism.

Finally, there is the problem already mentioned of a modern renewal of the ancient ethics of the good life as an ethics of existential self-realization. What is the point of this concern of personal ethics considered as a problem resulting from Kant's prioritizing of the universalization principle of justice over the search for my individual happiness? Let us at this point first come back to Foucault's verdict against the Kantian demand of a strictly universal ethics of the moral law in favor of a renewal of an ethics of the "*souci de soi.*"

It is an interesting fact, in this context, that Foucault, after his last public interview and shortly before his death, attended a conference on "human rights." And, so to speak, as an enlightened and progressive European intellectual, he—of course—pleaded for the concern of human rights. Here, obviously, an inconsistency in his position became visible, since the defense of human rights presupposes the universal validity of moral norms and wants to impose them to all states even on the level of international law.

Now, this inconsistency is rather characteristic for all cases, in which humanistic intellectuals want to defend

the uniqueness or even the incommensurability of individual or collective (i.e. socio-cultural) forms of life. Thus,
we find this attitude especially with cultural anthropologists and with post-modernist philosophers, e.g. with
Lyotard. They seem to plead exclusively for "difference,"
pluralism and, uniqueness; but, on a closer look, it is
revealed that it is universal acknowledgement of the rights
of being different or unique they are pleading for. Now, this
reflection shows that the usual confrontation of
(deontological) universalism and relativistic pluralism is in
fact a precipitous and superficial diagnosis of the actual (or
up-to-date) problem situation of ethics. What rather is indicated (suggested) is a situation of complimentarity between
the concern of diversity and uniqueness, on the one hand,
and that of universal validity of norms, on the other.

But this hint to a complimentarity situation is not yet a
solution of the moral problem that is indicated. For, there
is no principle of complimentarity from which we could
derive those unique forms of life we have to acknowledge
for ethical reasons. Or don't we need any moral reason for
this acknowledgement at all? Is the morally relevant
uniqueness of individual (and collective) forms of life
simply a matter of arbitrary decision or of an absolutely
free choice, as was suggested by the early Sartre? This
cannot be the case, if the free choice of a unique form of
life ought to be universally acknowledged as a human right.
There must be universally valid restrictions of the free
choice of a unique form of life that figure as necessary
conditions of a universal acknowledgement, e.g. demands
of justice. And at this point the Kantian principle of
universalization in a sense must again function as an
examination principle for the assessment of a unique
choice. The later Sartre recognized this, when he said

that everybody by choosing himself/herself must choose humanity. I shall come back to this proposal.

But even *a priori* restrictions placed on the choice of one's unique form of life don't provide a positive orientation for that choice. Here the point of the old question of the search for my good life, or my self-realization turns up again in a sharpened form. Autonomy of my reasonable will in Kant's sense certainly is a necessary condition of any morally acceptable choice of a form of life, but it is not yet a positive measure of my form of life yet to be chosen. Here rather the term *authenticity* points to the right direction, and this shows that Foucault at least had a point, when he played off the concept of *souci de soi* against the Kantian concept of the transcendental subject of reason.

But the following seems clear: with regard to the return, in sharpened form, of the old problem of the good life after Kant—a reappearance as the problem of one's authentic self-realization—we have so far only displayed a problem of a mediation of concerns that still calls for a solution by philosophical ethics. Is such a solution possible at all? Even such a question sounds immodest in our day, and for good reason.

For, as I intimated already in the preceding, we cannot solve the different ethical problems resulting after Kant's introduction of the universalization principle through deducting the solution from one principle. This means, I think, there is no metaphysical solution possible. It is neither possible to deduce the ought of ethics from a telos of being nor to deduce concrete normative orientations of one's good life from the Kantian principle of universalization. For this is indeed only a negative (restrictive) principle of examination with regard to possible maxims. But even as such the Kantian principle has not been strictly, i.e. transcendentally, grounded (justified), as Kant himself concedes at the beginning of the *Critique of Practical Reason* (1998), where he contents himself with declaring it to be an evident "fact of reason."

What Kant has provided instead of a transcendental foundation is rather a metaphysical explanation of the universalization principle by the "transcendent" ("intelligible") supposition of a "realm of purposes." But this account was part of the metaphysical doctrine of the "two realms" whose citizens human beings are said to be. Now, this dualistic conception rather prevented Kant from providing a plausible mediation between the ideal demands of reason and the empirical demands of pragmatic prudence, or of responsibility for the consequences of actions, or of authentic self-realization of individual persons in the context of their life as a whole. For with regard to the empirical orientation of life's practice Kant, strictly following his metaphysical dualism, had to suppose that human beings are no possible addressees of a moral ought at all, since they always are causally determined by their natural inclinations, and this was indeed what Kant presupposed in his famous postulate that a state of law must be constructed in such a way that it functions even for a people of devils, if only they are equipped with "intelligence." But if human beings are considered as members of the "realm of purposes," i.e. as autonomous co-subjects of reason, they would not need any "moral law" addressing them as an "ought."

Let us at this point try to introduce the transcendental-pragmatic approach of a foundation of dialogical or discourse ethics, which I consider as a possible basis for the solution of the post-Kantian problems outlined in the preceding.

3   The Transcendental-pragmatic Foundation of Dialogical or Discourse Ethics

Looked upon from the vantage point of a Kantian ethics the (transcendental-pragmatic) foundation of discourse ethics may be considered as a transformation of the Kantian approach after the linguistic-hermeneutic-pragmatic turn of philosophy in the last century, so to speak. In a sense, the basic structure of ethics, being a structure of reciprocity, is implied in the structure of communication by language. But this does not mean that we can ground ethics by extricating its basic structure from empirically given forms of language use or social communication, say by cultural anthropology, sociology, or socio-linguistics.

For, though we have good reasons for the presumption that the sought after basic structure of ethics is implied in the deep structure, so to speak, of all forms of human communication, we cannot directly derive the fundamental norms of ethics from given forms of life. For, let alone the impossibility of deriving an "ought" from an "is" or "fact of being," the given forms of life are never undistorted manifestations of those structures of communication that can be considered (to be) the basic structure of ethics. Even as substantial forms of morality (*Sittlichkeit* in Hegel's sense) they are always different compromises of the sought after (ideal) basic structures of communicative morality with other forms of human interaction—for example, strategic interaction—and, last not least, with the functional systems rationality of the political and economical institutions that necessarily belong to historically given socio-cultural forms of life. (Hence an attempt at grounding ethics as an ultimate normative basis for a "critical theory of society" cannot take recourse to the given forms of *Sittlichkeit* in the life world, since it must be capable of criticizing them in light (from the point of view) of an independent normative yardstick).

However, precisely that basic structure of ethics that must at least be implied in the deep structure of all human communication can be set free as an independent measure of social critique by strict reflection on the normative presuppositions of serious argumentative discourse about any topic whatsoever. By this transcendental-pragmatic reflection we can in a sense decipher what Kant could have meant by his talk of the evident "fact of (practical) reason." There are undeniable presuppositions of argumentation—such as equal rights and equal co-responsibility of all possible discourse partners—that constitute a "fact of reason" in the sense of an "*a priori* perfect." It is not a contingent fact of empirical knowledge, from which no moral norm can be derived, but a "*factum* of reason" in the sense of reason's having always already acknowledged those moral norms that belong to the necessary presuppositions or conditions of the possibility of argumentative discourse. We cannot deny or dispute these normative presuppositions of discourse without committing a performative self-contradiction, i.e. without destroying the "self-consistency of reason."

Thus, I cannot seriously argue in a discourse without forwarding universal truth-claims, i.e. claims to universal inter-subjective validity and hence claims to being possibly consented to by all members of an ideal discourse community; and this presupposes simultaneously, that all possible discourse partners have equal rights of judging my arguments; and, if the discourse is taken seriously, it must also presuppose that all partners are equally co-responsible for taking a stand to my arguments, and that means already: answering my questions. If somebody—like R. Rorty—says, philosophical discourse should be only something like "edifying conversation" and therefore did not presuppose universal validity claims, since one may content oneself with trying to persuade a particular audience, then either this very thesis as such performative contradicts itself, or it is no thesis at all and hence is irrelevant for the philosophical discourse, or—in the worst case—Rorty's statement is an attempt at immunizing his opinion against possible critique; for every serious critique presupposes universal validity-claims, and not just attempts at rhetorically persuading a particular audience.

The same counterarguments may be directed, for example, against Lyotard's *thesis*, that in a discussion we should not strive for consensus but for *dissensus*. And if Lyotard accounts for his self-contradictory thesis by the argument that striving for dissent furthers freedom and diversity of opinions and protects us against dogmatism and suppressive power, then he seems to presuppose that argumentative discourse can be equated to strategic negotiations where consent is forced through offers and threats, or to manipulative persuasions where autonomous consent to good reasons are *a priori* impossible.

But precisely these types of strategic rationality, which in our day—for example, in the economical theory of games—are often equated with rationality as such, precisely these modes of communication and interaction are excluded from the outset by serious argumentative discourse. We know from the outset that we cannot and must not use these ways of influencing other people in the service of our interests, if we want to find out the truth of a matter; for in this case we need the others—all possible members of an unlimited argumentation community—as possible helpers in proposing and criticizing validity claims and reasons for and against them.

Hence, the aim of consensus here cannot be pursued by suppression of possible dissent, but anyway only as a common concern to be striven for at the full risk of possible *dissensus*.

This shows that the *a priori* of argumentative discourse, which cannot be surpassed or replaced by anything else, when we want to solve a philosophical (or a scientific) problem, is indeed also the primordial insurance of human moral solidarity in a post-conventional sense. Hence, the possibility of a transcendental and rational foundation of ethics depends on the insight that the indubitable basis of our critical thought is not the Cartesian cogito, understood as the certainty point of "transcendental solipsism" in Husserl's sense, but it is the certainty point of public thought with a claim to inter-subjective validity, i.e. the point of arguing in a communication community.

Even Kant, who as a transcendental philosopher in the Critique of Pure Reason was attached to the cogito of the Cartesian tradition, when he tried to provide a foundation for ethics, found it necessary to introduce the conception of the intelligible "realm of purposes," which is nothing other than a metaphysical anticipation of the "ideal communication community" conceived of as a condition of possible consent about moral legislation. Now, it is this metaphysical prefigurement that is caught up by transcendental-pragmatic reflection on the necessary presuppositions of argumentation in a discourse; more precisely: it is the *a priori* of the ideal communication community and its normative presuppositions together with the history-dependent *a priori* of the real communication community, to which we belong as concrete persons, due to our socialization.

This exposition of the double *a priori* of the presuppositions of argumentative discourse brings us back to the problem of unfolding the approach of discourse ethics. For, in the preceding, we have only pointed to the moral presuppositions of any kind of argumentative discourse. But, a practical discourse of ethics comes about, if, or respectively, when not truth claims but moral rightness claims are raised and need to be settled by a practical discourse. In this case the rightness claims have to be discussed under the procedural conditions of an ideal discourse, that is, under the regulative idea of striving for a consent of all affected persons who could be conceived of as possible discourse partners. What does this mean?

In principle, all needs and interests of human beings that can become the subject of claims to moral rightness can and, if necessary, should become the topic of practical discourses, i.e. legal discourses as well as, ultimately, moral discourses; because these latter, although they may and must be supplemented and discharged by legal discourses, provide the last foundation also for the institution of law, as the discussion of human rights shows. From the principle of equal co-responsibility of all discourse partners it even follows that all human beings are somehow co-responsible for the application of discourse ethics with regard to the settlement of moral problems in the life world and, beyond that, even for the critical control, and the shaping and reshaping of such morally relevant institutions as, for example, the democratic state of law.

But I will not go into those problems in this chapter but concentrate on the problem of the relationship of discourse ethics and the post-Kantian concern of an ethics of the good life or self-realization, as I introduced them in the preceding.

4 Discourse Ethics and the Concern of an Ethics of Self-Realization

It seems clear that, on the transcendental-pragmatic account of discourse ethics, all morally relevant problems of the life world, and in this sense also the individual and collective concerns of striving for one's good life or self-realization, are subject to being judged, and, if possible, solved by practical discourses. (These practical discourses even include de-liberations that cannot be carried through publicly, but only *in foro interno*. For even these empirically solitary deliberations, according to the discourse principle, are through their inter-subjective validity claims *a priori* related to the forum of a public discourse, and that is, to the regulative idea of reaching a consensus of all affected persons. For this is the concretized substitute for the Kantian universalization principle. Anyway, also *in foro publico*, the interests of many affected persons always have to be represented by advocates.)

According to these conditions, also the individual and collective interests in a peculiar form of the good life have to be conceived of as intersubjective validity claims. This does not mean that, with regard to their material content, they are derived from, and imposed by, a law that is *a priori* valid for all, but it means that, through their conception as intersubjective validity claims, even as unique interests they become qualified as (human) rights (as was expressed already in the American constitution). But it also means that already as morally relevant interests they have to be mediated and thus restricted through universally valid norms of justice and co-responsibility in the face of common human problems.

45

This demand is today much more differentiated and problematic than it could have been in the days of Aristotle. For already the relationship between individual (personal) and collective concerns of the good life has become more problematic. In the Nicomachaean Ethics, Aristotle (1999) still made it clear that teaching ethics only makes sense in addressing mature persons who have been brought up "in a good *polis*." Thereby Aristotle still testified for the close and unproblematic relationship that in traditional societies tied together collective and individual (personal) forms of the good life. This changed already in the Hellenistic age, and today the Aristotelian testimony of the close interrelationship or interdependency between personal ethics and the collective ethics is either considered to be a characteristic trait of traditional societies or it is almost polemically confirmed by neo-Aristotelian communitarian philosophers. (Thus, for example, H.-G. Gadamer once tried to define what ethics still can be by the half Greek verdict: *HÔS DEI plus PHRONESIS*, i.e. what is proper and decent in one's society plus prudence).

But, under the characteristic conditions of modernity (including even the alleged post-modernity), i.e. under conditions of striving for authentic self-realization, as I supposed it as being the concern of Sartre as well as Foucault, there has arisen a remarkable difference and sometimes a strong tension between individual (personal) and traditional (collective) forms and concerns of the good life, although both concerns still are different from the strictly universalistic demands of a deontological ethics of justice. This means that the latter demands of ethics, which have claimed priority since Kant, and until J. Rawls, today has often the task of defending the individual (personal) concerns of the good life against the claims (demands) of traditional collective forms of the good life. This task has even become an important part of the world-wide engagement for human rights, especially in the case of the uncovering and defending women rights of self-realization as against traditional, culture-dependent forms of family structures and gender-relationships. (I think indeed that the terrible suppression of women rights by the Taliban in Afghanistan is much worse than the Chinese's suppression of political dissidents.)

Nevertheless, there is always also a more-or-less close internal connection between a particular, even unique form of striving for the good life as self-realization and a specific culture-dependent ethos-tradition. (This has been shown by cultural anthropology and by hermeneutic philosophy, and the conception of absolutely free choice or making of one-self, as was defended by the early Sartre, was simply untenable, as Sartre himself realized in his later writings.) We have even to recognize that belonging to a cultural ethos-tradition, e.g. to a religious community, is itself a human right that, if it is voluntarily chosen, is a precondition of personal self-realization. Hence, the ethical and juridical problematic of a multi-cultural society in our day gets its complexity by the fact that universalistic ethics of justice, and a law system that is committed to realizing human rights, is at the same time subjected to two obligations: to tolerate and protect the collective ways of the good life in the interests of their adherents and to defend the individual forms of self-realization against traditional or even fundamentalist claims of collective forms of ethnic or religious communities.

Now, all these tasks, on account of discourse ethics, cannot be fulfilled by a hierarchy of principles from which the concrete norms of public morality or law, or even the unique valuations of one's way of self-realization, could be deduced, say by a philosopher. The morally relevant solutions have to be reached through argumentative discourses of the affected persons or their advocates. And this is what is already going on in our day on the level of the thousand dialogues and conferences taking place within and beyond the national boundaries. It is going on in all those cases where the bearers of the different interests (individual and groups) are prepared to cooperate in a fair settlement of conflicts through discussion. I will not deal in this context with those cases, where the preconditions of practical discourses are not fulfilled, as e.g. in situations of war and civil war, or in those situations, where only strategic negotiations (through offers and threats) between power systems (also individual ones) are possible. But there is still another problem—the most delicate one—of the relationship between discourse ethics and an up-to-date ethics of authentic self-realization that has to be dealt with.

If we abstract from those cases where the free choice of one's way of the good life or self-realization has to be subjected to restrictions in favor of the universally relevant norms of justice or co-responsibility, then, as I suggested in the preceding, there remains the problem of choosing one's way of life; for example, of the choice of one's professional aims or of one's companion through life, etc., in short: the dimension of what Foucault called the *souci de soi*. In the preceding, I suggested already, that, in contradistinction to Kant's foundation of the principle of universalization, the unique choice of oneself is not primarily a concern of a person's autonomy as a reasonable being but rather a concern of his or her authenticity. Related to the validity-claims a person must have, in accordance to his or her communicative competence, especially in a dialogue with himself or herself *in foro interno*, it is not so much the claim to truth or rightness that is at stake, but the claim to truthfulness (or veracity, or sincereness). The way that has to be avoided is by deceiving oneself thoroughly and thereby missing the unique possibility and task of his or her life.

Now, here the question arises: is this dimension of ethical orientation completely outside the realm of discourse ethics—as it is indeed outside the dimension of what can be prescribed by a deontological ethics of justice? It seems to me that this cannot be supposed, if one thinks through the transcendental-pragmatic foundation of discourse ethics, i.e. if one takes into consideration that one needs to make understandable one's choice to oneself and to others. The realm of trying to give reasons in a discourse reaches so far the realm of the logos as hermeneutical logos, so to speak. But this means, in my opinion, that also a certain dimension of the claims to universal, intersubjective validity remains active. This does not mean that the uniqueness or the incommensurability of the individual (personal) dimensions of authenticity gets lost, that it could be derived from universal principles. But it seems, nevertheless, to mean that, in choosing the unique way of ourselves, we must avoid falling victim to arbitrariness, that we must try to remain in connection with what can be recognized as being valid for every reasonable being.

This, I suggest, was Sartre's point in *L'existentialisme est un humanisme* (1996), when he corrected his former position of groundless freedom by the verdict that we can and must "choose humanity by choosing ourselves." Indeed he comes close to the quasi-Kantian postulate that everybody who were exactly in my situation in every respect, would have to make the same choice as I have to make. Of course this would only be a "regulative idea," to which, according to Kant ,"nothing empirical can ever correspond." Nevertheless, it builds a bridge between the ethics of authentic self-realization and the ethics of justice as universalization.

At the conclusion of this chapter, let me express the following conjecture (presumption): From my presupposition that—contrary to the predominant opinion of our time—there is indeed a strong tension but not an absolute contradiction between the universalistic discourse ethics and the modern version of the ethics of the *souci de soi*, it also follows that an internal connection between the ethical problematic of authentic self-realization and the cognitive interest of the hermeneutic and critical science of psychoanalysis, or more generally expressed, deep psychology, is conceivable (as has been suggested by the Neo-Frankfurt School of Critical Theory and also by P. Ricœur some decades ago). The internal connection between the universalistic (autonomous) logos of discourse-rationality, i.e. of giving reasons in a discourse or dialogue, and the hermeneutic logos of an attempted understanding of the unique authenticity of self-realization and of its possible failures by self-deception makes it possible, I suggest, that there is such a thing as the project of people's emancipation from hidden (i.e. unconscious) constraints, a project that connects ethics and the concern of deep psychological sciences.

[4]

Emotion, Metaphor and the Evolution of the Mind

Arnold H. Modell

Department of Psychiatry, Harvard Medical School
Cambridge, MA, USA

1. The modern brain is a mosaic structure containing vestiges of earlier homologous designs to which are added newer cognitive devices that have radically altered its organization. So that when one approaches the problem of the evolution of the mind, continuities and discontinuities need to be identified. The juxtaposition of older and more ancient primate-like functions alongside what is uniquely human is especially apparent when we compare emotions in animals and man. Darwin noted that lower animals are excited as we are by the same primordial emotions of fear and rage. Intelligent mammals such as monkeys and dogs share with us what he called more complex emotions, such as jealousy and curiosity. Dogs as well as human beings desire to love and be loved, from which it may be inferred that evolution has preserved homologous brain structures that generate the expression of emotion.

2. There is recent evidence to support this inference in that it appears that the same anatomical structures mediate fear in both rats and humans. Joseph LeDoux (1996; 2002) has shown that the emotion of fear in rats is mediated by the amygdala. Recent experiments using the non-invasive technique of magnetic resonance scans (fMRI) on human subjects has shown that the amygdala is also specifically activated as a response to the exposure of fearful faces. It is probable that the amygdala activates fearful responses in all mammalian species. But lest we take this homology too far, we should remind ourselves that in all other mammals including primates the reaction to a fearful stimulus is immediate and involuntary, where we, in certain contexts, have the capacity to experience fear in small doses which can be pleasurable; rats do not enjoy going to horror movies.

3. This invariant and involuntary emotional activation is something that we share with other species and yet at the same time transcend. In considering this problem of discontinuity, it is useful to think in terms of the degree of freedom from internal and external environmental inputs, which our unique linguistic capacities provide. For example, chimpanzees, our genetically closest neighbor, when emotionally aroused cannot suppress their vocal cries. Jane Goodall (1968) writes: "Chimpanzee vocalizations are closely tied to emotion. The production of a sound in the absence of the appropriate emotional state seems to be an almost impossible task for a chimpanzee." Goodall goes on to describe that "On one occasion when Figan (a chimpanzee at the Gombe Stream Reservation) was an adolescent, he waited in camp until the senior males had left and we were able to give him some bananas (he had none before). His excited food calls quickly brought the big males racing back and Figan lost his fruit. A few days later he waited behind again, and once more received his bananas. He made no loud sounds, but the calls could be heard deep in his throat almost causing him to gag."

4. It can be said that Figan, although he remembered the past, was bound to the present. This observation accords well with Gerald Edelman's distinction between primary and higher order consciousness. Primary consciousness is the remembered present; perceptual inputs evoke specific categorical memories; primary consciousness can be described as episodic, a sequence of separate events, strung together like beads in a necklace. Whereas higher order consciousness is a many layered consciousness that enables the individual to create a model of past, present and future, thus freeing one from the tyranny of ongoing events. This schema or internal model of past, present and future provides a sense of continuity and coherence which could be described as the biological self. The sense of self as an organizing, coherent-making and meaning generating agency, is either absent in primates or present in only a very rudimentary form.

5. The linguist Derek Bickerton (1995) has proposed a similar idea. He contrasts two basic modes of thinking that he calls on-line thinking and off-line thinking. On-line thinking focuses on the immediate environment. He defines on-line thinking as computations carried out only in terms of neural responses elicited by the presence of external objects, whereas off-line thinking involves computations carried out on more lasting internal representation of those objects. Bickerton also believes that some primates and dolphins have a proto-language, but it is a language that lacks syntactic structures that does not enable them to go off-line. For Bickerton, the discontinuity between ourselves and other species, that which makes us uniquely human, is our generative grammar that allows us to go off-line.

6. There is considerable controversy regarding the evolution of language and the adaptive advantage that language provides—is language necessary for thought or can one think using only visual images? Did our ancestors acquire the use of symbols to create meaning before acquiring syntax, or was it the other way around? Some linguists believe that a mutation led to the relatively sudden acquisition of syntax, while some psychologists as well as neuro-scientists such as Edelman believe that semantics, the attribution of meaning, preceded the acquisition of syntax. I tend to favor this latter view and will hypothesize that in the evolution of language metaphor has played a decisive role in the cognitive revolution that preceded the appearance of spoken language.

Metaphor can be defined as the mapping of one conceptual domain onto a dissimilar conceptual domain. The term derives from the Greek verb *metaphorein,* to transport or transfer. Metaphor is not simply a figure of speech, a departure from literal meaning, it is the invisible engine behind language and thought. I have come to believe that metaphor is the currency of mind. It is by means of metaphor that we generate new perceptions of the world, and it is through metaphor that we organize and make sense out of experience. Metaphor is a cognitive tool that is essential for thought. Metaphor no less than language derives from a neuro-physiological process. This is evident in the fact that when we dream we generate involuntary visual metaphors. This observation did not escape Darwin who referred to the dream as "an involuntary art of poetry."

7. Darwin wrote (1977): "The mental powers of some early progenitor of man must have been more highly developed than any existing ape, before even the most imperfect form of speech could have come into use." Darwin proposed that a cognitive revolution had to occur before language developed. I will suggest that metaphor may have played a salient role in that cognitive revolution.

8. I shall propose that there are two broad classes of metaphor: fixed or frozen metaphors and generative metaphors. Frozen metaphors are perceptual metaphors that link affectively charged memories of the past, especially the memory of trauma, to current perceptual inputs. A metaphoric correspondence between past and present will greatly enlarge the class or category of which that memory is a member. Traumatic memories, which carry a high affective charge, can be evoked by a metonymic association (a part substituting for the whole) with a current perceptual element—provided, however, that the metonymic element is embedded in a perception that corresponds metaphorically to the traumatic memory. It is an evident adaptive advantage to enlarge through metaphoric correspondence, the field of vigilante. In psychiatry, certain symptoms, such as phobias and inhibitions, can be understood as derivatives of frozen metaphors in which there is an invariant correspondence between a past categorical memory and current perceptual inputs. Transference phenomena in which the emotional memory of a salient relationship in the past is projected onto someone in the present, can also be understood as a derivative of a fixed or frozen metaphor. Again, a metonymic association to an aspect in the person in the here and now will evoke the affective memories of the past. In transference and traumatic memories, the perceptual metaphoric correspondence between the past and the here and now is unambiguous, that is to say, it is relatively fixed or frozen. On the other hand, generative metaphors, are ambiguous conceptual metaphors wedded to language and are generative in the sense that new meanings are created. In the transfer of concepts from one domain to an entirely different domain, there is a space between the metaphor and the thing "metaphorized" that allows for the selection of individually created meanings. Such metaphors require language, whereas frozen metaphors require only an internal schema or image. Here are some examples of generative conceptual metaphors: "The poet is the priest of the invisible; the first hour of the morning is the rudder of the day; sex is the poor man's opera."

9. Researchers believe that modern speech and modern language appeared between 50,000 and 100,000 years ago. Darwin's supposition that man's progenitors developed higher mental powers before the advent of even imperfect speech is consistent with endocasts of fossil skulls. Changes were observed in those areas of the brain that mediate language functions such as a growing asymmetry of the brain in small brain hominids. The development of a communicative capacity greater than the higher apes is suggested by the appearance of a true homo-like Broca's area and organizational changes in the posterior parietal cortex. It is possible that these changes in the parietal area facilitated intermodal connections essential for metaphor. The ground for metaphor was probably made fertile by the development of a normative synesthesia, where equivalent meanings are registered across different sensory modalities.

10. That these changes occurred prior to the appearance of modern speech is supported by the research of the linguist Philip Lieberman (1984), who claims that our capacity to produce the sounds of human speech depends upon the descent of the vocal apparatus within the neck. Chimpanzees, who have no voluntary control over their respiration, have their vocal apparatus at the back of the pharynx. Lieberman believes that the vocal apparatus of archaic hominids as well as Neanderthals do not have a fully descended vocal apparatus, although there is some controversy regarding his assertion.

11. Although Neanderthal's brains were larger than ours, they are thought to have had a non-human vocal tract. Their speech if present, Lieberman believes, was imprecise and less rapid as compared to our own. Therefore archaic hominids who showed left brain asymmetry, and lack the vocal apparatus for modern speech, presumably would have the capacity for a more complex method of communication as compared to primates, but a capacity that fell far short of modern speech. If archaic humans did not have the apparatus for modern speech, how did they think and how did they communicate?

12. We know something of the mind of modern apes and we know even more about our own minds, but what lies in between us and our nearest primate relatives is speculative. The fossil record indicates that there is a total absence of cultural artifacts until the late Paleolithic period approximately 30–40,000 years ago. In the preceding millennia our hominid ancestors improved the quality of stone tools but there was no attempt at stylistic embellishment or other artifacts suggesting the presence of mythic thought. Regarding cultural artifacts, there was sharp break in the archaeological record, providing evidence of a creative explosion that occurred about 30,000 years ago in the late Paleolithic period. There is no question that these artifacts were produced by a race that was anatomically identical to ourselves.

13. It was once thought that Neanderthals practiced a burial ritual, and that they decorated their bodies with ochre. If this proves to be true it would be of enormous significance for it would mean that their minds were capable of mythic thought. But some scholars now discount this possibility as all the Neanderthal fossils were found in caves; it seems as if the burials were not intentional but accidental. So the fact that the Neanderthals presumed belief in an after life has not been proven. Although there is no evidence pointing to a mythic Neanderthal culture, there is, however a recent finding of a musical instrument constructed from a bone dated between 43 and 67 thousand years ago.

14. The Neanderthals and before them, Homo Erectus had a sufficient social organization to migrate widely from their origins in Africa into such inhospitable northern ranges as Siberia. Recent reports, not yet fully confirmed, indicate that some archaic species of man migrated into Siberia as early as 300,000 years ago. To sustain themselves in such a cold climate would require some degree of social planning and social organization to ensure a constancy of food supplies. If modern speech and modern language had not yet evolved, how then did these people communicate with each other? An attractive hypothesis has been offered by the psychologist Merlin Donald (1993). He proposed that our ancestors evolved through three distinct cognitive stages, described as episodic, mimetic and mythic. With each advance in cognition there were corresponding changes in memory systems. The episodic stage, characteristic of higher primates, is equivalent to Edelman's remembered present where the individual's response is bound to associations evoked by events in current time. The primate's life is tied to the memory of concrete episodes; the past dictates the response to the here and now.

15. As I mentioned, there is some evidence from fossil endocasts that brain asymmetry and increased complexity of the left parietal cortex appeared before the evolution of modern speech and language, suggesting that our ancestors had a method of communication more complex than primate communication yet falling short of modern speech. Donald proposed that the social organization of this middle period was based on mimetic gestures. The use of mimetic gesture does not of course preclude vocalizations and singing. Through fixed or frozen metaphor, mimetic gestures could acquire an expanded semantic content. Although the fossil record of this period indicates no tangible cultural artifacts, other than the manufacture of stone tools, it is plausible that such mimetic metaphoric gestures when combined with rhythmic movements and song began to assume the nature of ritual. In this fashion language development may have been semantically bootstrapped prior to the acquisition of syntax. Thought and communication based upon fixed metaphor endowed gestures with meaning prior to the acquisition of a generative grammar. Furthermore, Donald proposed that each cognitive advance was accompanied by a more complex system of memory. A social organization based on mimetic gesture would support collective memory, which in later millennia would evolve into an oral tradition. This cognitive advance would have radically altered the relation of the present to the past.

16. The cultural revolution of the late Paleolithic was evidenced in the creation of cave paintings and rock sculptures including moveable small sculptures such as the so-called Venus figurines. This cultural explosion required modern speech and language and generative conceptual metaphors. From the cave paintings we can infer the presence of an oral tradition and a highly developed mythic culture. There is here a sharp discontinuity with earlier ancient peoples. What is remarkable is the uniformity of the paintings throughout Western Europe which suggests a shared social organization extending over a very extensive geographical area. The archaeological evidence indicates that there is nothing to suggest a comparable social organization amongst the Neanderthals or Archaic Homo Sapiens. While we have not been able to decipher the symbolic meaning of these beautiful cave paintings, there is no doubt that they served a religious function as they were found in nearly inaccessible dark recesses of limestone caves. The paintings are almost entirely of animals captured in motion. With relatively few species chosen for representation, such as horses, oxen, reindeer, bison, lions, bears and rhinoceroses. As one expert, Leroi-Gourhan (1964) commented, there must have been an oral context behind the symbolic assemblage of figures. The paintings illustrate a religious parable, the script of which has been lost.

17. Language and mythic culture therefore represents the clearest discontinuity between us, our primate ancestors and our nearest human progenitors. As I noted earlier, when we compare the expression of emotion in animals and humans, primates are always on-line, that is, they are unable to delay or suppress a response to an emotionally evocative stimulus. Our affective responses are also on-line, in the sense that they are automatic and involuntary, like the new born infant's cry when separated from its mother. But we also learn to regulate and control our affective responses. With regard to the expression of emotion it is as if we posses both an ancient, primate-like brain and a modern mindful brain. How might the evolution of mythic culture effect the expression of emotion? Let us imagine a group of our ancestors who are particularly sensitive to loud noises and become terrified in thunderstorms. So that the sound of thunder evokes uncontrolled fear and flight. But then let us imagine that their culture teaches them that when it thunders the gods are angry and are hurling stones at each other. Thunderstorms then do not evoke uncontrolled flight as the myth provides an explanation for that which is otherwise terrifying. All cultures have myths that explain the origin of life and what happens after death. Myth provides an explanation concerning certain universal fears such as the fear of death and the terror of being alone in the universe. It can be said that mythic beliefs are embodied, they are grounded in the body, in primitive fears and what can be called an epistemic impulse, a universal urge to make sense out of one's experience. Through myth we enter into another level of consciousness that removes us from the here and now. This expansion of consciousness provides a measure of freedom from environmental inputs. If we can explain the experience of emotion it does not matter whether we attribute our emotions, to the action of Gods, or to the action of our unconscious, or to our limbic system. The act of explanation itself introduces another level of consciousness which enables one to delay an immediate response to environmental inputs.

18. Myth opens another level of consciousness that allows us to go off-line. The metaphoric transfer of affective experience in the here and now into another domain allows the individual a greater measure of voluntary control over affective responses. This does not mean, however, that in mythic cultures there was a belief that emotions are attributed to the agency of the self. The attribution of the control of emotions to the self is a late development of Western culture.

19. The psychologist Julian Jaynes (1976) proposed a theory that in an earlier stage in the evolution of consciousness, emotions were thought to be placed within the individual by the Gods, so that the individual was controlled by forces outside of the self, but controlled nevertheless. According to Jaynes' hypothesis a mythic or a collective group mind substituted for the agency of self. For instance in Homer's Iliad there are many examples of the Gods placing emotions in the minds of mortals and thereby controlling their actions. When Agamemnon robs Achilles of his mistress, it is a God that grasps Achilles by his yellow hair and warns him not to strike Agamemnon. Myth provides for what can be described as a doubling of consciousness—events occur both in the here and now and simultaneously in an entirely separate domain: the realm of the imaginary. The mythic narrative can be thought of as a metaphoric organizing of experience at another level of consciousness. It is this multi-leveled consciousness that allows us to go off-line.

20. Amongst the many unanswered questions of paleo-archaeology, there is the question: why was there an absence of novelty in ancient societies? Until the advent of Homo Sapiens, the manufacture of artifacts such as stone tools remained relatively unchanged for millennia. In this respect archaic Homo Sapiens, and Neanderthals were no different from chimpanzees despite their large brains. One can attribute the introduction of novelty to language, which allows one not only to represent reality but to also transform it. But the introduction of novelty must still be attributed to an individual and be understood in accordance with the psychology of the individual. There must be a space within language and culture for the idiosyncratic interpretation.

21. Open metaphors because of their ambiguity offer such possibilities. The novelist Walker Percy in discussing metaphor, observed that there is a space between the name and the thing that allows the individual to make mistakes in understanding. It is these mistakes or purely personal apprehensions that introduce a measure of freedom and create new forms of understanding. To state it another way, metaphor opens the door of the imagination. And it is imagination and not only language that makes us uniquely human.

[5]

ETHOS IN ACTION

Franco F. Orsucci

Institute for Complexity Studies &
Institute of Psychiatry and Clinical Psychology
Catholic University of Rome, Italy
franco.orsucci@ixtu.org

> *The Way floats and drifts*
> Lao-Tzu

1   Evolution

During his lecture of September 9, 2000, in Zürich, Jurgen Habermas was moving from Kant's theory of justice and Kierkegaard's theory of self-being to defend the right of abstention in crucial questions for the good or "non-failed" life. His paper on the "Risks of liberal genetics" (2001; 2003) takes a firm position without renouncing his post-metaphysical thinking.

He reminds us that Adorno's *Minima Moralia* (1974) starts by just citing the *Gaia Scientia* of Nietzsche (1994), and admitting a failure: "the *Gaia Scientia* was considered for such a long time as the proper field of philosophy . . . the doctrine of the right way," but this way, Adorno assumes, is somehow lost because in the meanwhile ethics has regressed to the state of "sad science." It can produce just dispersed and aphoristic "meditations on offended life." Adorno's moral reflections throw the dark light of pessimism on the shadows of society, but lack an attempt to establish positive possible solutions.

As long as philosophy has been believing in its capacity to provide a general framework of right and good life for subjects and communities, this kind of life would have been a universal model. Universal in the sense that world religions see in their respective founder's life an ideal path of rescue. In this sense, philosophy, or a large part of it, was trying to build a religion without gods.

Now ethics tends to circumscribe its normative action to justice and, when it is applied to rights and dues of community or individuals, it changes. The questions of right and good become context-dependent and intermingled with interrogations on the identity of the speaker, considered as the object-and-subject of discourse. In the end, following this approach, we tend to rely on situation ethics: a system of ethics by which acts are judged within their contexts instead of following categorical principles.

In this post-modern drift, psychotherapies risk surrogating ethics, by orienting lives or simply dispensing consolations. This tendency might also be related to the always latent social attitude of confusing mental disturbance and ethical disorder; the kind of confusion neither good for ethics, nor for psychotherapies.

Kierkegaard could be considered the first to offer an answer to this post-modern problem by proposing the post-metaphysical concept of "being-able-to-be-yourself." But, we are confronted, nowadays, with a sudden change in the boundaries and the sense of self (and non-self): the by-product of a new evolutionary jump for the human species.

Adults are going to consider as producible and mouldable the genetic matter of their children and plan its design. They might exert on their genetic products a power of disposition; the power to penetrate the somatic base of spontaneous reference and ethical freedom of another person, hitherto considered possible just on things, not on persons.

In this way the boundary between human and non-human is going to change. If and when this change is accomplished, children will consider, for example, their parents, as creators of their genome, responsible for the desired or undesired consequences.

This possible new scenario derives from a change in the social distinction between things and persons, subject and object, and poses new questions:

- How does the self-understanding of a genetically programmed person change?
- How does this change the areas of creativity, autonomy, free-will and equality in human relations?
- Can we foresee a "right to a genetic heritage non-compromised by artificial operations?"

What in Kant was the reign of necessity becomes the reign of disposability.

These questions are going to bring to light the similarity between ethics and the immune system: both are the guardians of the Self at the social or at the body level. They define the boundaries between self and non-self in different but correlated domains.

Perhaps we are all (the growing numbers that have entered into the sphere of this transference) "*les commencements d'une mutation.*" We can see it: all of us in the near future being described as the early stages of a mankind where alterity and intimacy have been expanded to the point of recursive interpenetration.

2 Local Rules

At the first Artificial Life Workshop (Langton, 1989) Craig Reynolds presented a computer simulation of bird flocking based on three simple rules (Langton, 1991). Each virtual bird was required to:

1. Maintain a minimum of distance from other objects, including other birds;
2. Match velocities with neighboring birds;
3. Move toward the perceived centre of birds' mass in the neighborhood.

These rules are all local, applying to individuals, and yet their effect was that of flocking dynamics of striking realism and elegance.

Flocking is here an emergent global phenomenon: it is interesting how the three rules capture the essence of the standard three-value set [Liberty (1), Equality (2), and Fraternity (3)] basic to French society. It is worth noting that this value set is based just on simple contextual, local rules about relation dynamics.

Of special interest is the way in which the component elements emerging during such self-organization are both mutually constraining and mutually sustaining. Each is a vital local part of the global pattern within the bounded space. A different pattern can be engendered within the same space and with the same values, but the significance is distributed into different clusters, whether differently located, of a different size, or of a different number.

Another basic feature of this experiment is that computer simulations based on Artificial Life can show the birth of ethics and values in some sort of *in vitro* experiments. The history of humanity has shown enough about the risks of *in vivo* experiments in ethics and politics, and this seems an opportunity to test ethical hypotheses before any "collateral damage."

These experiments highlight the structuring process of value sets in a *small world*, by applying just local and not universal rules. Organization of values is the emergent production of local iterated interactions. This kind of contextual organization seems more flexible and adaptable to unexpected events than "top-down" rigid organizations.

From such a perspective it is clearly far less appropriate to attempt to focus on any particular pattern of values. Of much greater relevance is recognizing the process whereby different kinds of contextual circumstances can evoke the emergence of such different patterns from the value space. This small-world process is evidence-based during its development. Results are observable during each step of the process, while they are not so transparent in universalistic approaches.

Another way to look at such patterning is in terms of the "pathways in ethical landscapes" that may emerge between different value locations. Just as a pattern of mountain valleys may severely condition the nature of relationships between otherwise proximate zones, particular values may also affect (or be dependent on) each other to a greater or lesser degree.

The cultural historian William Irwin Thompson synthesizes: "Values are not objects, they are relationships" (Thompson, 1996). As Gregory Bateson stated: "Destroy the pattern which connects and you destroy all quality" (Bateson, 1979). This means that values, just as *qualia*, are (embedded in) the mesh of relations.

3 Human Rights

We are searching for a reliable foundation for the existing convergence of these two basic principles, the good and the right, private and public, in different combinations (see De Risio, this volume). The difficulties on this path could be appreciated in the modern renewal of the ancient ethics of the good life as an ethics of self-realization.

In this context, Michel Foucault, after his last public interview and shortly before his death, attended a conference on "Human Rights." And, so to speak, as an enlightened and progressive European intellectual, he pleaded for the concern of human rights. Here, obviously, an inconsistency in his position became visible, since the defence of human rights presupposes the *universal* validity of moral norms. The logical consequence is that Foucault suggested solaces as the basic form of ethical practice (Foucault, 1978; 1988).

Now, this inconsistency is rather typical for all cases in which humanistic intellectuals try to defend the uniqueness or even the alterity of individual or collective (i.e. socio-cultural) forms of life. Thus we find this attitude especially with cultural anthropologists and with post-modernist philosophers, e.g. with Lyotard. They seem to plead exclusively for difference, pluralism and, uniqueness; but, on a closer look, it is revealed that it is the universal acknowledgement of *universal rights to be different or unique* they are pleading for.

The usual confrontation of universalism and relativistic pluralism is in fact a precipitous and superficial diagnosis of the current situation of ethics. Rather, what is suggested is a complimentarity between the concern of diversity and uniqueness and the universal validity of norms.

There must be some universally valid restrictions as necessary conditions of the universal acknowledgement of life (see Apel, this volume).

4 Language and Consciousness

In the "Nicomachaean ethics," Aristotle (Aristotle, 1999) asserted that teaching ethics only makes sense in addressing mature persons who have been brought up "in a good polis." But, what does "teaching ethics" mean? Thereby Aristotle still testified for the close and unproblematic relationship that in traditional societies tied together collective and individual (personal) forms of the good life. Mediation between private and public ethics was made by tradition. This changed in the Hellenistic age, when some sort of mass society began, and today the Aristotelian position of the close interrelationship or interdependency between personal ethics and collective ethics is considered to be a characteristic trait of traditional societies.

The dissolution of tradition in contemporary global societies generates a need for the definition of a new medium and matrix.

Language reveals itself as the matrix and mediation of relations: the basic medium for ethics. In language games people negotiate the interactions of their individual forms of life by accepting, at the same time, the constraints of shared linguistic codes. The basic structure of ethics, being a structure of reciprocity, is implied in the structure of communication in language. This does not simply mean that we could ground ethics by extricating its basic structure from empirically given forms of language use or social communication.

Following this path, we need to supersede both the representationalist approach to language and the idealist generative grammar: both of these approaches are seriously challenged by advancements in the scientific studies on nonlinear dynamics of language and neurosciences.

In this new perspective, ethical thinking should be considered mostly as a subsystem of the semiotic universe, apt to preserve and maintain the boundaries of the (individual and social) Self—a function that is quite similar to the functioning of the immune system: preserving the psychobiological self by discerning what is Self and what is Non-Self; a semiotic agent exploring through confusions and intrusions along their fuzzy boundaries. Bioethics works on the edge between body, mind, and society, just in the same area of functioning of the immune system: in the fuzzy and transitional areas between private and public. There you have to accept the "private language paradox" *à la* Wittgenstein (Kripke, 1982; Wittgenstein, 1967) just to conjugate the reasons of body and mind.

The paper written by Francisco Varela about his personal experience of organ transplant is paradigmatic on this.

> I found myself spontaneously desiring a reciprocity, to seal a pact with the anonymous donor. (. . .) This translates on the imaginary level to the presence of the donor in the gift itself, attached to it, and following its transferences.   (Varela, 2001)

5 Time and Subject for Action

The birth of bioethics might throw more light on the nature of human subjectivity, between the immediate copying of bodies and the reflections of social institutions. We have seen that language games, based on small-world rules, can mediate ethical processes. Now we are going to focus on timing and action, using the lenses of contemporary neurosciences to highlight the complex textures that form the fabric of the ethical subject.

St. Augustine in the Confessions (specifically book II) writes about the paradoxes of "a past-containing nowness" (*dislenlio unimi*) (Augustine, 1990). More recently, the pragmatic psychology of William James, as found in the famous *Principles of Psychology* (1920), proposes the definition of *specious present*.

Time in experience is quite different from time as measured by a clock. Time in experience presents itself not only as linear but also as having a *complex texture* (evidence that we are not dealing with a "knife-edge" present), a texture that dominates our existence to an important degree (Varela in Petitot, 1999).

As phenomenological research and ecological psychology (Gibson, 1979) has repeatedly emphasized, *perception is based in the active interdependence of sensation and movement. It is this active side of perception that gives temporality its roots in living.* Within this general framework, we will concentrate more precisely on the structural basis and consequences of this *sensory-motor integration* for our understanding of temporality.

This overall approach to cognition is based on *situated embodied agents.* Varela (Varela, 1991; Thompson, 2001) has introduced the name *enactive* to designate this approach more precisely. It comprises two complementary aspects:

(i) the ongoing coupling of the cognitive agent, a permanent coping that is fundamentally mediated by sensory-motor activities;

(ii) the autonomous activities of the agent whose identity is based on emerging, endogenous configurations (or self-organizing patterns) of neuronal activity.

From an enactive viewpoint, any mental act is characterized by the concurrent participation of several functionally distinct and topographically distributed regions of the brain and their sensory-motor embodiment. From the point of view of the neuroscientist, it is the complex task of relating and integrating these different components that is at the root of temporality.

These various components require *a frame or window of simultaneity that corresponds to the duration of lived present*. There are three scales of duration to understand the temporal horizon just introduced:

- basic or elementary events (the "1/10" scale);
- relaxation time for large-scale integration (the "1" scale);
- descriptive-narrative assessments (the "10" scale).

The first level is already evident in the so-called *fusion interval* of various sensory systems: the minimum distance needed for two stimuli to be perceived as non-simultaneous, a threshold that varies with each sensory modality. These thresholds can be grounded in the intrinsic cellular rhythms of neuronal discharges, and in the temporal summation capacities of synaptic integration. These events fall within a range of 10 msec (e.g. the rhythms of bursting inter-neurons) to 100 msec (e.g. the duration of an EPSP/IPSP sequence in a cortical pyramidal neuron). These values are the basis for the 1/10 scale.

Behaviorally, these elementary events give rise to *micro-cognitive phenomena* variously studied as perceptual moments, central oscillations, iconic memory, excitability cycles, and subjective time quanta. For instance, under minimum stationary conditions, reaction time or oculomotor behavior displays a multimodal distribution with a 30–40 msec distance between peaks; in average daylight, apparent motion (or "psi-phenomenon") requires 100 msec.

This leads naturally to the second scale, that of *long-range integration*. Component processes already have a short duration, of the order of 30–100 msec. How can such experimental psychological and neurobiological results be understood at the level of a normal cognitive operation? A long-standing tradition in neuroscience looks at the neuronal bases of cognitive acts (perception-action, memory, motivation, and the like) in terms of cell assemblies or neuronal ensembles. A cell assembly (CA) is a distributed subset of neurons with strong reciprocal connections.

The diagram depicts these processes. A cognitive activity (such as the head turning) takes place within a relatively incompressible duration, a *cognitive present*. The basis for this emergent behavior is the recruitment of widely distributed neuronal ensembles through increased frequency, coherence in the gamma (30–80 Hz) band. Thus, the corresponding neural correlates of a cognitive act can be depicted as a synchronous neural hypergraph of brain regions undergoing bifurcations of phase transitions from one cognitive present content to another.

Recently, this view has been supported by widespread findings of oscillations and synchronies in the gamma range (30–80 Hz) in neuronal groups during perceptual tasks.

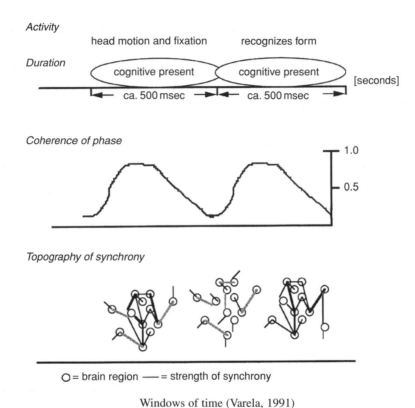

O = brain region ——— = strength of synchrony

Windows of time (Varela, 1991)

Thus, we have neuronal-level constitutive events that have a duration on the 1/10 scale, forming aggregates that manifest as *incompressible but complete cognitive acts* on the 1 scale. This completion time is the basis for the origin of duration without an external or internally ticking clock.

Nowness, in this perspective, *is therefore pre-semantic* in that it does not require a rememoration in order to emerge. The evidence for this important conclusion comes, again, from many sources. For instance, subjects can estimate durations of up to *2–3 seconds* quite precisely, but their performance decreases considerably for longer times; spontaneous speech in many languages is organized such that utterances last 2–3 seconds; short intentional movements (such as self-initiated arm motions) are embedded within windows of this same duration. The *Ur-impression* is the proper mode of the now, or in other words, it is where the new appears; impression addresses the new. Briefly: *impression is always presentational, while memory or evocation is representational.*

This brings to the fore the third duration, the 10 scale, of *descriptive-narrative* assessments. In fact, it is quite evident that these endogenous, dynamic horizons can be, in turn, linked together to form a broader temporal horizon. This temporal scale is inseparable from our descriptive-narrative assessments. It constitutes the "narrative center of gravity" in Dennett's metaphor (Dennett, 1991), *the flow of time related to personal identity.* It is the continuity of the self that breaks down under intoxication or in pathologies such as schizophrenia or Korsakoff's syndrome. As Husserl (1980) points out, commenting on similar reasoning in Brentano: "We could not speak of a temporal succession of tones if . . . what is earlier would have vanished without a trace and only what is momentarily sensed would be given to our apprehension." To the appearance of the just-now one correlates two modes of understanding and examination:

(i) remembrance or evocative memory;
(ii) mental imagery and fantasy.

This (ethical) behavior embodies the important *role of order parameters in dynamical accounts.* Order parameters can be described under two main aspects:

(i) The current state of the oscillators and their coupling, or initial conditions; and
(ii) The boundary conditions that shape the action at the global level; the contextual setting of the task

performed, and the independent modulations arising from the contextual setting where the action occurs (namely, new stimuli or endogenous changes in motivation).

6 Doublings

Around 1911 Husserl introduced the term *double intentionality* (1980) for this articulation, since there is not only retention (of the object-event) but also retention of retention (a reflective awareness of that experience). James came to a similar conclusion when he described his experience of lying in bed, telling himself to get up. He stayed in bed. He later found himself up but could not remember the exact instant of transition. Libet (1993; 1999; Freeman, 2001) experiments on the consciousness lag, though discussed in their methodology, suggest the same solution: there is a doubling between intentionality and consciousness, and it has neurobiological foundations.

Vitiello (2001), starting from the "quantum brain" hypothesis (whose problems we won't discuss here) suggests: "In the dissipative quantum model of brain memory recording is modelled as coherent condensation of certain quanta in the brain ground state. The formation of finite size correlated domains allows the organization of stored information into hierarchical structures according to the different life-times of memories and the size of the corresponding domains. The openness of the brain to the external world (dissipation) *implies the doubling of the brain system degrees of freedom.* The system obtained by doubling, the *Double, plays the role of the bath or environment in which the brain is permanently embedded. It is suggested that conscious as well as unconscious activity may find its root in the permanent dialogue of the brain with its Double."* (present author's Italics).

Jacques Lacan (1977) schematized this dialectics in the *mirror phase* of infant development: the birth of the subject's identity in front of the mirror is also the moment in which the *Ur-spaltung* between the subject of intentionality, the Es, and the subject of consciousness, the Ego, crystallizes.

The doubling dynamics that such different scholars have highlighted, though coming from very different perspectives and suggesting different explications about their nature, can be represented in a diagram, the *"I scheme"* (Orsucci, 2002):

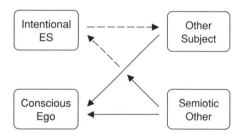

This scheme about the *dynamics in subjective experience* offers the crossing of two main vectors: the **vector of the imaginary,** between OS and the Ego; and the **vector of the unconscious** experiences, between the (Semiotic) Other and the Es. Ethics in action is located just at this crossroads.

7   Intention and Narration

The neurodynamics of time we have been pursuing is essentially based on nonlinear coupled oscillators: constitutional instabilities are the norm and not a nuisance to be avoided (Varela, in Petitot, 1999). The case of *Gestalt* perceptual multi-stability makes this quite evident experientially: the percepts flip from one to another (depending on order parameters) by the very nature of the geometry of the phase space and its trajectories.

Top: a highly schematic diagram of the current view that a complex dynamic should be regarded as having a geometry in phase space where *multiple instabilities* are found locally (gray lines). A system's trajectory shifts (black lines)

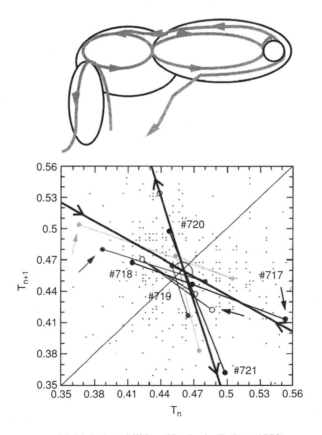

Multiple instabilities (Varela, in Petitot, 1999)

constantly from one local instability to another, in an unceasing flow, under the modulation of boundary conditions and initial conditions. Bottom: experimental evidence for local saddle instabilities in the time series from the cortex of an implanted epileptic patient. The peaks of the local discharge in the temporo-occipital cortex are followed in their return map or Poincaré section.

"Even the most precise consciousness of which we are capable is affected by itself or given to itself. The very word consciousness has no meaning apart from this duality" (Merleau-Ponty, 1962). Alterity is the primary clue for time's constitution. We are affected not only by representations and immanent affection ("*affection de soi par soi*"), but by the inseparable alterity from the sphere of an *ego-self*. We could add that these dualities generate the *multiple instabilities of consciousness.*

The very distinction between auto and hetero ceases to be relevant, since in all cases it all comes down to the same manifestation: it is a question of *something other*, the experience of an alterity is constitutive in the paradoxical nature of the shifter called Ego—subject and object at the same time, or with an infinitesimal delay.

A possible non-dual synthesis would be located where *affect* is constitutive of the self and at the same time contains a radical openness or unexpectedness concerning its occurring. This bootstrap principle seems to be present in a variety of natural systems and has recently been referred to as *operating at the edge of chaos*, or *self-organized criticality*. This idea provides a renewed view of evolution, since it provides an answer to the old nature (genetic expression) versus nurture (environmental coupling conditions) dilemma. In this synthetic view (Kauffman, 1993) the relation between natural forms and the selection process in their ecological embedding is not one of contradiction but of precisely mutual imbrications. This built-in shiftiness, enfolding trajectories and geometry, gives a natural system a possibility of always staying close to regions in phase space that have multiple resources.

8 The Ethical Thing

The subject of ethics has a ambiguous, but vital and creative, nature built on the imbrications between the immediate copying of bodies and the reflections of social institutions.

This dynamical nature is embedded in the intra subjective dialectics between presentation and representation, intentionality and reflection. That is a result of the repeated meetings of doubles. There is a "readiness-for-action, a micro-identity and its corresponding level, a micro-world . . . we embody stream of recurrent micro-world transitions" (Varela, 1991). And these micro worlds are nested and imbricated in other worlds.

Somewhere, there is a small time-lag, like a butterfly-wings' flap, that generates through multiple cascades the doubling and mirroring effect that we call consciousness.

This constitution of the subject founds the status of decision, volition and action. In presentation and action there is an embedding of representation and imagination (Lakoff and Johnson, 1999).

The emergence of ethics is the extremely complex process including in its stream the cascade of all these events. Ethics emerges in language when the subject in action meets the things or *Das Ding* (Lacan, 1966; 1992), the Real, because in ethics there is always an index fingering a possible action.

The word of ethics is not symbolic *per se*: it is a dialogue with pure presence.

BIBLIOGRAPHY

Adorno T.W. (1978) *Minima Moralia, Reflections from Damaged Life*, London: NLB.

Apel K.-O. (2001) The Response of Discourse Ethics to the Moral Challenge of the Human Situation as such and Especially Today (Mercier lectures, Louvain-la-Neuve, March 1999), Leuven: Peeters.

Aristotle (1999) *Nicomachaean Ethics*, Indianapolis, In: Hackett Pub. Co.

Augustine & Augustinian Heritage Institute (1990) *The Works of Saint Augustine*, Brooklyn, NY: New City Press.

Baker N. (1996) *The Size of Thoughts: Essays and Other Lumber*, New York: Random House.

Bateson G. (1979) *Mind and Nature a Necessary Unity*, New York: Dutton.

Bickerton D. (1995) *Language and Human Behavior*, Seattle: University of Washington Press.

Cabanne P. (1971) *Dialogues with Marcel Duchamp* (translated by Ron Padgett), New York: Viking Press.

Darwin C. (1977) *The Collected Papers of Charles Darwin*, Chicago: University of Chicago Press.

Dawkins R. (1982) *The Extended Phenotype*, Oxford and San Francisco: Freeman.

Dennett D. (1984) *Elbow Room*, Cambridge, Mass.: MIT Press.

Dennett D. (1991) *Consciousness Explained* (1st edn.), Boston: Little, Brown and Co.

Dennett D. (1991) *Consciousness Explained*, Boston: Little, Brown and Co.

Dennett D. (1995) *Darwin's Dangerous Idea, Evolution and the Meanings of Life*, New York: Simon & Schuster.

Bibliography

Dennett D. (1998) *Brainchildren*, Cambridge, Mass.: MIT Press.

Diamond J. (1997) *Guns, Germs and Steel*, New York: Norton.

Donald M. (1993) *Origins of the Modern Mind: Three Stages in the Evolution of Culture and Cognition*, Cambridge, Mass.: Harvard University Press.

Foucault M. (1973) *Madness and Civilization: A history of Insanity in the Age of Reason*, New York: Vintage Books.

Foucault M. (1978) *The History of Sexuality*, New York: Pantheon Books.

Foucault M. (1988) *Technologies of the Self: A Seminar with Michel Foucault*, Amherst: University of Massachusetts Press.

Freeman W.J. (2001) *How Brains Make up Their Minds*, New York: Columbia University Press.

Gibson J.J. (1979) *The Ecological Approach to Visual Perception*, Boston: Houghton Mifflin.

Goodall J. (1968) *The Behaviour of Free-living Chimpanzees in the Gombe Stream Reserve*, London: Bailliáere, Tindall & Cassell.

Guston P. (Bricker Balken D.) (2001) *Philip Guston's Poor Richard*, Chicago: University of Chicago Press.

Habermas J. (2001) Die Zukunft der menschlichen Natur, auf dem Weg zu einer liberalen Eugenik?, Frankfurt am Main: Suhrkamp.

Habermas J. (2003) *The Future of Human Nature*, Malden, Mass.: Polity.

Husserl E. (1980) *Collected Works*. Boston: The Hague.

James W. (1967) *The Writings of William James: A Comprehensive Edition*, New York: Random House.

Jaynes J. (1976) *The Origin of Consciousness in the Breakdown of the Bicameral Mind*, Boston: Houghton Mifflin.

Kant I. (eds. Cassirer H.W., Heath King G., and Weitzman R.) (1998) *Critique of Practical Reason*, Milwaukee, Wis: Marquette University Press.

Kant I. (ed. Gregor M.J.) (1998) *Groundwork of the Metaphysics of Morals*, Cambridge, U.K.: Cambridge University Press.

Kant I. (eds. Guyer P. and Wood A.W.) (1998) *Critique of Pure Reason*, Cambridge: Cambridge University Press.

Kant I. (eds. Lauchlan Heath P. and Schneewind J.B.) (1997) *Lectures on Ethics*, New York: Cambridge University Press.

Kauffman S.A. (1993) *The Origins of Order, Self Organization and Selection in Evolution*, New York: Oxford University Press.

Kitcher P. (1997) *The Lives to Come the Genetic Revolution and Human Possibilities*, New York: Simon & Schuster.

Koestler A. (1964) *The Act of Creation*, New York: Dell.

Kripke S.A. (1982) *Wittgenstein on Rules and Private Language*, Cambridge, Mass.: Harvard University Press.

Lacan J. (1966) *Écrits*, Paris: Editions du Seuil.

Lacan J. (1992) *The Ethics of Psychoanalysis, 1959–1960*. New York: Norton.

Bibliography

Lakoff G. and Johnson M. (1999) *Philosophy in the Flesh, the Embodied Mind and its Challenge to Western Thought*, New York: Basic Books.

Langton C.G. (1989) *Artificial Life: Proceedings of an Interdisciplinary Workshop on the Synthesis and Simulation of Living Systems*, September 1987, in Los Alamos, New Mexico. (v. 6th edn.) Redwood City, Cal.: Addison-Wesley.

LeDoux J.E. (1996) *The Emotional Brain: The Mysterious Underpinnings of Emotional Life*, New York: Simon & Schuster,.

LeDoux J.E. (2002) *Synaptic Self, How our Brains Become Who We Are*, New York: Viking.

Leroi-Gourhan A. (1964) Le Geste et la Parole, Paris: A. Michel.

Libet B., Freeman A. and Sutherland K. (1999) *The Volitional Brain: Towards a Neuroscience of Free Will*, Thorverton, U.K.: Imprint Academic.

Marcuse H. (1962) *Eros and Civilization. A Philosophical Inquiry into Freud*, New York: Vintage Books.

Merleau-Ponty M. (1962) *Phenomenology of Perception*, New York: Humanities Press.

Nietzsche F.W. (1994) *The Complete Works of Friedrich Nietzsche*, Stanford, Cal.: Stanford University Press.

Orsucci F. (2002) *Changing Mind: Transitions in Natural and Artificial Environments*, Singapore: World Scientific Publishing.

Orsucci F., Ed. (1998) *The Complex Matters of the Mind*, Singapore: World Scientific.

Petitot P., Ed. (1999) *Naturalizing Phenomenology, Issues in Contemporary Phenomenology and Cognitive Science*, Stanford, Cal.: Stanford University Press, 1999.

Philip Lieberman. (1984) *The Biology and Evolution of Language*, Cambridge, Mass.: Harvard University Press.

Poincare H. (ed. Halsted G.B.) (1929) *The Foundations of Science: Science and Hypothesis, The value of Science, Science and Method*, New York: The Science press.

Poundstone W. (1985) *The Recursive Universe: Cosmic Complexity and the Limits of Scientific Knowledge*, New York: Wm Morrow.

Sartre J.P. (1996) L'existentialisme est un Humanisme, Paris: Gallimard.

Smith J.M. and Szathmary E. (1995) *The Major Transitions in Evolution*, Oxford: Freeman.

Thompson E. and Varela F.J. (2001) Radical Embodiment: Neural Dynamics and Consciousness, *Trends Cog Sci.* 5 (10):418–425.

Thompson W.I. (1996) *Coming into Being: Artifacts and Texts in the Evolution of Consciousness*, New York: St. Martin's Press.

Valery P. (1974) (ed. Robinson J.). *Cahiers*, Paris: Gallimard.

Bibliography

Varela F.J., Thompson E., and Rosch E. (1991) *The Embodied Mind: Cognitive Science and Human Experience*, Cambridge, Mass.: MIT Press.

Vitiello G. (2001) *My Double Unveiled: the Dissipative Quantum Model of Brain*, Amsterdam: John Benjamins Pub. Co.

Wittgenstein L. (1967) *Philosophical Investigations*, Oxford: Oxford University Press.

INDEX

Index

Index